D1451882

ANIMAL SCIENCE
Biology & Technology
THIRD EDITION

Robert Mikesell

MeeCee Baker

Delmar Cengage Learning
is proud to support
FFA activities

Join us on the web at

agriculture.delmar.cengage.com

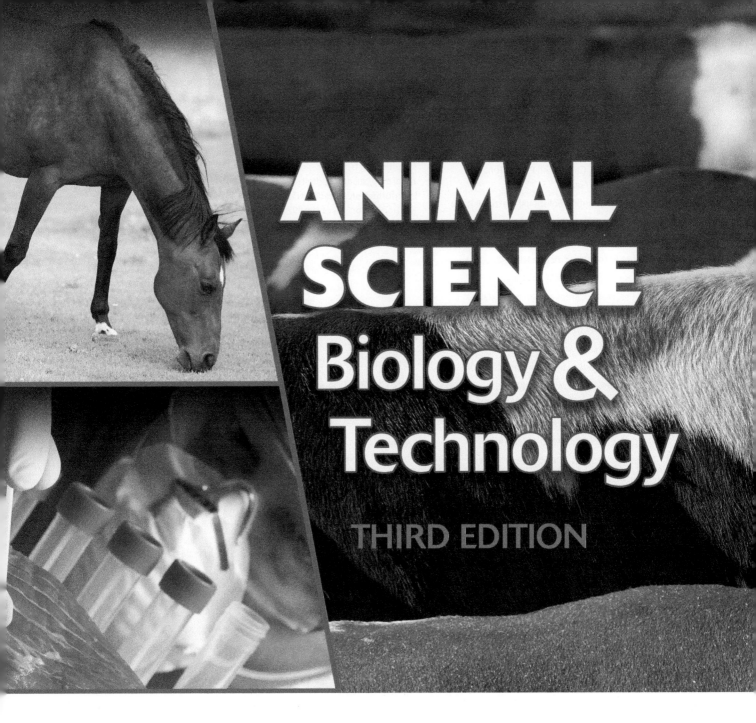

ANIMAL SCIENCE
Biology & Technology
THIRD EDITION

ROBERT MIKESELL, PhD
Senior Instructor
Department of Dairy and Animal Science
The Pennsylvania State University
University Park, Pennsylvania

MEECEE BAKER, PhD
Partner/Chief Operating Officer
Versant Strategies
Harrisburg, Pennsylvania

DELMAR
CENGAGE Learning

Australia • Brazil • Japan • Korea • Mexico • Singapore • Spain • United Kingdom • United States

Animal Science Biology and Technology, Third Edition
Robert Mikesell and MeeCee Baker

Vice President, Career and Professional Editorial: Dave Garza

Director of Learning Solutions: Matthew Kane

Acquisitions Editor: Benjamin Penner

Managing Editor: Marah Bellegarde

Product Manager: Christina Gifford

Editorial Assistant: Scott Royael

Vice President, Career and Professional Marketing: Jennifer Baker

Marketing Director: Debbie Yarnell

Marketing Manager: Erin Brennan

Marketing Coordinator: Jonathan Sheehan

Production Director: Carolyn Miller

Production Manager: Andrew Crouth

Content Project Manager: Katie Wachtl

Senior Art Director: David Arsenault

Technology Project Manager: Tom Smith

Production Technology Analyst: Thomas Stover

For product information and technology assistance, contact us at
Cengage Learning Customer & Sales Support, 1-800-354-9706

For permission to use material from this text or product,
submit all requests online at **www.cengage.com/permissions.**
Further permissions questions can be e-mailed to
permissionrequest@cengage.com

Library of Congress Control Number: 2010923212

ISBN-13: 978-1-4354-8637-9

ISBN-10: 1-4354-8637-4

Delmar
5 Maxwell Drive
Clifton Park, NY 12065-2919
USA

Cengage Learning is a leading provider of customized learning solutions with office locations around the globe, including Singapore, the United Kingdom, Australia, Mexico, Brazil, and Japan. Locate your local office at:
international.cengage.com/region

Cengage Learning products are represented in Canada by Nelson Education, Ltd.

To learn more about Delmar, visit **www.cengage.com/delmar**

Purchase any of our products at your local college store or at our preferred online store **www.CengageBrain.com**

Notice to the Reader

Publisher does not warrant or guarantee any of the products described herein or perform any independent analysis in connection with any of the product information contained herein. Publisher does not assume, and expressly disclaims, any obligation to obtain and include information other than that provided to it by the manufacturer. The reader is expressly warned to consider and adopt all safety precautions that might be indicated by the activities described herein and to avoid all potential hazards. By following the instructions contained herein, the reader willingly assumes all risks in connection with such instructions. The publisher makes no representations or warranties of any kind, including but not limited to, the warranties of fitness for particular purpose or merchantability, nor are any such representations implied with respect to the material set forth herein, and the publisher takes no responsibility with respect to such material. The publisher shall not be liable for any special, consequential, or exemplary damages resulting, in whole or part, from the readers' use of, or reliance upon, this material.

This book was previously published by Pearson Education, Inc.

Printed in the United States of America
2 3 4 5 6 7 8 21 20 19 18 17

CONTENTS

Introduction   / vi
Preface   / vii
About the Authors   / x
Acknowledgments   / xi
Design image Credits   / xii

UNIT I
Physiology

Chapter 1   Cellular Biology and Animal Taxonomy   / 2
Chapter 2   Biology of Growth and Development   / 20
Chapter 3   Muscle and Meat Biology   / 38
Chapter 4   Biology of Digestion   / 50
Chapter 5   Biology of Reproduction   / 66
Chapter 6   Genetics   / 86
Chapter 7   Ethology: Animal Behavior and Welfare   / 102

UNIT II
Application

Chapter 8   Swine Management and the Swine Industry   / 118
Chapter 9   Beef Cattle Management and the Beef Industry   / 160
Chapter 10   Dairy Cattle Management and the Dairy Industry   / 196
Chapter 11   Sheep Management and the Sheep Industry   / 228
Chapter 12   Horse Management and the Horse Industry   / 258

UNIT III
Evaluation

Chapter 13   Judging Contests   / 296
Chapter 14   Performance Data   / 312
Chapter 15   Meats Judging   / 330
Chapter 16   Livestock Judging   / 366
Chapter 17   Dairy Cattle Judging   / 406

Glossary   / 428
Bibliography   / 436
Index   / 439

INTRODUCTION

By Dr. Baker

I taught agriculture.

I needed an animal science text.

The text needed to build a strong foundation of scientific principles, incorporate applied management practices, review the industry based on recent data, and emphasize selection.

In agriculture teacher terms, after using the text, I wanted my students to be able to explain the physiology behind cattle heat cycles, raise a market hog, select an animal science career based on industry outlook, and deliver a sound set of reasons. This book was written based on those needs.

Animal Science Biology and Technology was written with the agricultural student in mind. Broad in scope, the text remains detailed enough to assist the student with an agriscience project, Supervised Agricultural Experience program, or judging contest. Moreover, this text approaches the discipline of animal science from a new perspective. Unit I, physiology, addresses cellular function, growth and development, digestion, reproduction, genetics, and ethology. Unit II, application, teaches the practical uses of animal science. This segment is organized by species: swine, beef cattle, dairy cattle, sheep, and horses, and provides details of recent industry trends and issues. Unit III instructs students in the interpretation of performance data for livestock evaluation as well as judging of meat, swine, sheep, and beef and dairy cattle.

Attractively illustrated, this comprehensive text will appeal to any student interested in production livestock. Each chapter in the text begins with objectives, key terms, and an introductory paragraph. Chapters also include summaries, terms to define, evaluation questions, experiential learning opportunities, and career investigations. The ancillary material includes a laboratory manual with teacher's instructions, instructor's guide, PowerPoint slides, and a comprehensive test question bank. The ClassMaster CD-ROM makes utilizing the supplementary materials easy for time-stressed instructors.

PREFACE

Although not commonly glamorized in the media, animal agriculture is a critical component to the economy of the United States. This total economic impact comes not only from the production of livestock but from the feed, transportation, processing, distribution, and marketing of animal products as well. Obviously many careers are tied to production of meat, milk, fiber, and the horse industry.

Less than 2 percent of America's population is directly involved in any phase of production agriculture. However, individuals employed in associated industries must understand basic livestock physiology, the application of this physiology, and the industries surrounding each specie, as well as the evaluation of live animals.

To address these needs, this text is arranged in three units. In Unit I, physiology, students will gain an understanding of basic cell structure as well as the various systems within an animal's body. Many of these concepts apply to human physiology as well. The last chapter in this section explains concepts of animal behavior and welfare—skills necessary for anyone working with animals.

Unit II reveals the application of the concepts learned in the physiology segment. Each specie of livestock is examined from breeds to feeding programs to housing and common health concerns. Additionally, this unit discusses the background of the industry pertaining to each specie. Livestock producers constantly deal with ever-changing biological, economic, and regulatory hurdles to operating successful enterprises. Those interested in livestock should be aware of these issues and able to discuss them intelligently.

In Unit III, students will prepare to compete in judging contests for swine, beef, sheep, and dairy cattle, as well as meats. Many FFA and 4-H members compete in these contests each year. Judging contests are valuable tools not only for recognizing differences among individual animals, carcasses, or cuts of meat but also for learning decision-making and communications skills.

The authors hope this text is comprehensive enough to give students a thorough insight into the livestock industry yet simple enough for the uninitiated to comprehend.

EXTENSIVE TEACHING/ LEARNING PACKAGE

INSTRUCTOR'S GUIDE

Each chapter of the Instructor's Guide includes a helpful guide for creating lesson plans, a list of chapter objectives, the Chapter Motivator, an Introduction to the material to be covered, chapter term matching exercise with answer key, answer key to the review questions, and a summary.

LAB MANUAL

The lab manual for *Animal Science Biology and Technology* is designed to help the instructor provide practical application to the technical material found in the text.

LAB MANUAL INSTRUCTOR'S GUIDE

The lab manual instructor's guide for *Animal Science Biology and Technology* is designed to help the instructor provide answers and direction for the laboratory activities, which support the technical material found in the text.

CLASSMASTER CD-ROM

This supplement provides the instructor with valuable resources to simplify the planning and implementation of the instructional program. It includes customizable test bank questions in the ExamView program, instructor slide presentations in PowerPoint, a pdf of the Instructor's Guide, and an extensive Image Library.

ABOUT THE AUTHORS

The authors of *Animal Science Biology and Technology* are Robert Mikesell and MeeCee Baker.

Dr. Mikesell, lead author, earned his bachelor's, master's, and PhD degrees in Animal Science from The Pennsylvania State University. He currently serves as a senior instructor in Penn State's Department of Dairy and Animal Science where he teaches livestock judging, integrated animal biology, and value determination of meat animals.

Dr. Baker taught agriculture for over 20 years prior to being appointed to the executive office of the Pennsylvania Department of Agriculture. More recently, Baker became a partner in Versant Strategies, an agricultural public affairs firm. She earned B.S. and PhD degrees in Agricultural Education from The Pennsylvania State University. Her M.S. degree in Agricultural Economics was completed at The University of Delaware. She was the first female to be elected NVATA President in 1996.

In addition, the authors have experience in raising, exhibiting, and judging livestock.

ACKNOWLEDGMENTS

The authors of *Animal Science Biology and Technology* want to thank the Delmar Cengage Learning Team for affording them the opportunity to republish this text and providing the expertise necessary to bring the project to completion.

The authors and Delmar Cengage Learning would also like to thank the following individuals for their participation in the review process:

Chris Wilder
Williston High School
Williston, FL

Tracy Champagne
South Fork High School
Stuart, FL

DESIGN IMAGE CREDITS

COVER

Horse: © William Murphy/iStockphoto
Analysis of animal material: © dra_schwartz/iStockphoto
Crowded little pigs: © Cristi Matei/Shutterstock
Cows background: © Yuriy Chaban/Shutterstock

DESIGN

Unit 1 Tab: ©Jeanne Hatch. Used under license from
Shutterstock.com
Unit 2 Tab: Photo courtesy of Robert Meinen
Unit 3 Tab: Photo courtesy of Penn State College of
Agricultural Sciences

Front Matter Corner Tabs:

© Jean Scheigen

© Stock.Xchng

© Andrea Kratsenberg

© Michael Schat

Minor Column Design:

Cow © Samual Rosa

Sheep © Sarah Joos

Horse © Stock.Xchng

Career Focus: © David Chasey
Animal Science Facts: © Stock.Xchng

Back Matter Corner Tabs:

© Lali Masriera

© Keran McKenzie

© Art Explosion

Physiology

1

Cellular Biology and Animal Taxonomy

CELLS—SMALL BUT MIGHTY

Objectives

After completing this chapter, students should be able to:

▶ Draw and label the cellular structures

▶ Describe the functions of cellular structures

▶ Differentiate between mitosis and meiosis

▶ Discuss the relationship of tissues, organs, and systems

▶ Explain how taxonomy is used to classify living organisms

▶ Use binomial nomenclature to write scientific names of common livestock

▶ Trace the domestication of farm animals

Key Terms

apoptosis	endoplasmic reticulum	meiosis
binomial nomenclature	feral	mitochondria
cell	gametes	mitosis
cell membrane	genes	nucleus
centromere	golgi bodies	ribosomes
chromosomes	haploid	taxonomy
cytoplasm	Linnaeus	
DNA	lysosomes	

Career Focus

Jill was fascinated with cells ever since she first saw a single-celled bacterium under a microscope. Growing up in rural Texas, Jill also liked to raise and exhibit her FFA cattle. She decided to pursue a career in the animal sciences with an emphasis on learning about cells and cellular functions. During her collegiate career, she worked in a laboratory on campus that concentrated on how muscle cells repaired themselves after being injured. Jill learned the intricate ways these specialized cells detected the need for repair, how genes were notified that repairs were necessary, and how those genes created new proteins to repair damaged muscle tissue. Her undergraduate job prepared her for a fantastic career in medical research where she now investigates ways to increase the rate of muscle repair in humans with a goal of reducing the time it takes to heal muscle injuries.

INTRODUCTION

The basis for understanding animal science biology and technology lies in understanding the **cell** (smallest biological unit). Far from being simple, the animal cell is complex and performs many different functions. Cells take in nutrients, excrete waste products, secrete proteins, perform cellular work, respond to their environment, and reproduce themselves. Organisms vary from single-celled bacteria to familiar, multicellular whole animals, such as swine, sheep, cattle, and horses.

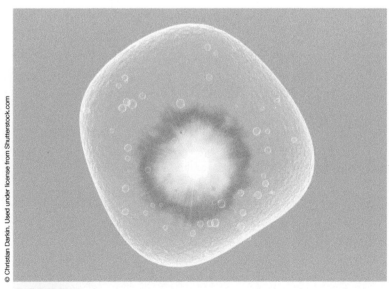

© Christian Darkin. Used under license from Shutterstock.com

FIGURE 1-1

A cell is the basic unit of life.

CELLULAR STRUCTURE AND FUNCTION

Cell theory states that a cell is the basic unit of life and that a cell only originates from other cells. All organisms are made of one or more cells. Functions of all cells are similar in the following ways:

1. They must take up nutrients from their external environment.

2. They must excrete waste products into their external environment.

3. They must do some kind of work, such as synthesize proteins (example, liver cells), store energy (example, fat cells), carry oxygen (example, red blood cells), transport electrical impulses (example, nerve cells), store minerals (example, bone cells), or move (example, muscle cells).

4. They must reproduce themselves.

Cells can be considered to be little factories with different departments responsible for varying duties. The outside wall of the factory is called the **cell membrane**. The cell membrane is made of a thin layer of lipid (fat-like film) that separates the cell contents from the external environment. Imbedded in this lipid layer are specialized protein doors that allow large molecules (raw materials, such as carbohydrates or proteins) to pass into the cell. Newly made proteins (finished products) and cellular waste products pass out of the cell through these protein doors. Individual organelles, or compartments, within the cell are also separated by lipid layers, like walls would separate rooms within a manufacturing plant.

Inside the cell factory is the office, or **nucleus**, which is surrounded by the nuclear membrane. The nucleus controls all cell activity. **Chromosomes** are small strands of genetic material that reside in the nucleus. They are made of **DNA** (Deoxyribonucleic Acid)—a genetic compound that controls inheritance. See figure 1-2. Chromosomes contain many small,

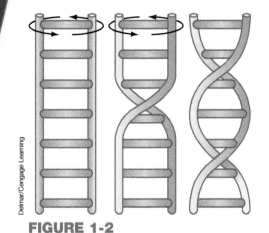

Delmar/Cengage Learning

FIGURE 1-2

British chemist Rosalind Franklin first described the spiral shape of DNA in 1952.

coded pieces of DNA called **genes**. Chromosomal genes contain the blueprint for the work the cell has to do as well as instructions for the cell to replicate itself. Genes control specific economically important traits as well as physical characteristics. Even though every cell contains all the genes for an entire organism, only those needed to do the certain type of work that the cell is destined to accomplish are active. For example, a liver cell has the genes to become a heart, but only the genes needed to perform liver functions are active in a liver cell.

The jellylike substance between the cell membrane and the nucleus is called **cytoplasm**. Within the cytoplasm are the various departments of the cell factory, which are collectively called organelles.

Raw materials entering the cell are lipids, carbohydrates, and proteins. The cell often modifies the raw materials into useful forms so the cell can feed itself or make products for secretion. The **endoplasmic reticulum**, a network of membranes that connects the cell membrane to the nucleus, manufactures, processes, and transports all incoming raw materials and outgoing cellular products. New proteins are manufactured by **ribosomes**, many of which are attached to the endoplasmic reticulum. The **mitochondria**, small, egg-shaped organelles, manufacture adenosine triphosphate (ATP), which is used as an energy source for the cell. **Lysosomes** are round organelles whose function is to digest and recycle molecules that are no longer useful. Cellular products undergo final assembly and packaging in the cellular factory through a series of flat, membrane-encased **golgi bodies**. See figure 1-3.

CELLULAR REPRODUCTION

Cells go through a life cycle of growth and reproduction called the cell cycle. The entire cell cycle has two phases: interphase (the period between cell divisions) and cell division. Most of a cell's life is spent in interphase. Interphase is further divided into three periods: G1, S, and G2. During the first period of

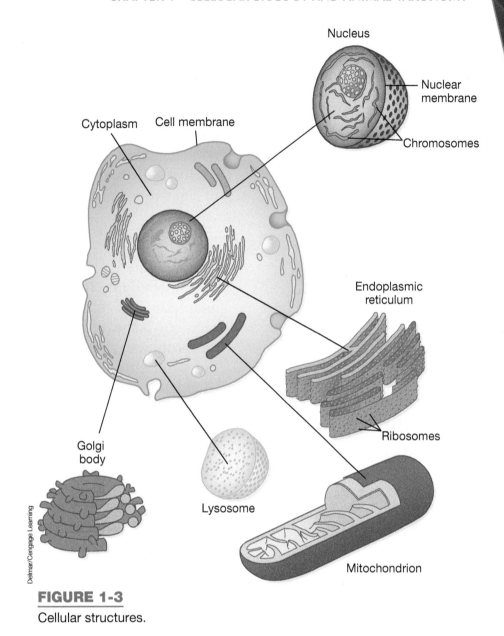

FIGURE 1-3

Cellular structures.

interphase (G1), the cell grows in size by increasing the number of organelles and the volume of cytoplasm. During the S period, or second period of interphase, the genetic material replicates or copies itself so that there are two identical sets of chromosomes called sister chromatids. At this point in the cycle, chromatids are still attached to each other at a central point called the **centromere**. During the third period of interphase, called the G2 period, the cell manufactures organelles and prepares for cell division, or **mitosis**.

At several points during the cell cycle, the cell inspects the DNA replication process to determine if genetic material has been properly copied. If the cell discovers an error in the copying process, it destroys itself through a process called **apoptosis** (programmed cell death). Failure of a cell to identify improperly copied genetic material and further replication of imperfect cells is a step in the development and growth of cancerous tumors.

Mitosis

Mitosis, the actual division of non-sex cells, is further divided into four periods: prophase, metaphase, anaphase, and telophase. At the conclusion of these phases, what was formerly one cell becomes two. During prophase, the nuclear membrane disappears and the sister chromatids shorten and thicken into identifiable chromosomes. During metaphase, the sister chromatids line up along the central axis or equator of the cell. One half of each pair of sister chromatids attaches to spindle fibers, which pull one set of chromosomes to each of the two new daughter cells. During anaphase, the chromatids separate, each moving toward opposite ends of the dividing cell. Each of the two new daughter cells forms a new nuclear membrane during telophase. Cytokinesis occurs when the cytoplasm and organelles evenly divide between the two new cells and a new membrane forms between them. See figure 1-4.

Meiosis

Animal sex cells (ova and sperm), or **gametes**, divide a little differently than other cells in a process called **meiosis** or reduction division. To examine the difference, we must understand chromosome number and what happens to chromosomes at fertilization.

Each species of animal has a certain number of chromosomes present in duplicate in every cell of its body. This number is called the **haploid** or N number. For example, the haploid number for humans is 23. Because each chromosome is present in duplicate, the

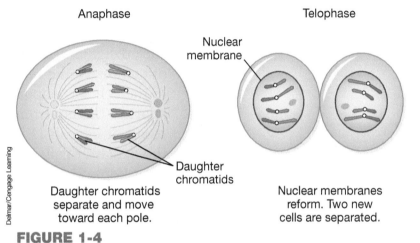

Prophase

Nuclear envelope
Centromere

Nucleolus

Chromosomes visible.
Nuclear membrane fragments.

Metaphase

Equator   Spindle fibers

Replicated chromosomes are
aligned at equator. One-half
of each pair of sister chromatids
attach to spindle fibers.

Anaphase

Daughter
chromatids

Daughter chromatids
separate and move
toward each pole.

Telophase

Nuclear
membrane

Nuclear membranes
reform. Two new
cells are separated.

Delmar/Cengage Learning

FIGURE 1-4

Mitosis is the division of non-sex cells.

total number of chromosomes is twice the haploid number. This number is referred to as the diploid or 2N number. Therefore every human cell has 46 total chromosomes or 23 identical or homologous pairs of chromosomes. The process of meiosis reduces sex cells from 2N to N. So, when ova and sperm (gametes) meet and combine at fertilization, the resulting embryo will have 2N chromosomes, 1N from each of its parents. Therefore, animals inherit genes and physical characteristics from both parents.

The process of meiosis is divided into two steps: meiosis I and meiosis II. Both steps are similar to mitosis, except the end result is four haploid cells instead of two diploid cells.

Meiosis I begins with prophase. As with mitosis, the chromosomes thicken and become visible. Chromosomes are present in homologous pairs with a total of four sister chromatids per pair. During metaphase, the homologous pairs line up on the axis of the dividing cell opposite from each other. During anaphase, the homologous pairs of chromosomes leave each other and are pulled toward opposite poles by spindle fibers. During telophase, the cells physically divide. Each daughter cell now contains one chromosome from each pair. Meiosis I is called reduction division because the number of chromosomes is reduced from diploid to haploid. Remember that during mitosis both daughter cells are diploid.

During the second portion of meiosis or meiosis II, each daughter cell again divides. During metaphase II of meiosis II, the sister chromatids line up on the axis of the cell. In anaphase II, the chromatids are pulled apart at the centromere by spindle fibers. The end result of meiosis II is two haploid cells from each of the daughter cells exiting meiosis I. Each of the four final haploid cells or gametes contains one strand from the original homologous pair of chromosomes. See figure 1-5.

CELLS, TISSUES, ORGANS, AND SYSTEMS

In multi-cellular animals, individual cells are specialized to perform a specific task. For instance, muscle cells are responsible for support and locomotion. Bone cells make up the skeleton and are responsible for structural support. Red blood cells carry oxygen. Fat cells store energy but do not carry oxygen. Some specialized types of cells combine to make up tissues. For example, cartilage, which is found in the human nose and ear, is a type of flexible connective tissue that acts as a shock absorber for the body.

Cells and tissues work together to form organs, such as the liver, which manufacture substances vital for body function.

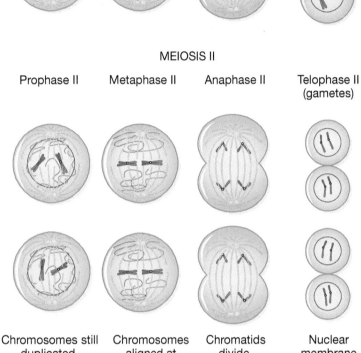

MEIOSIS I

Prophase I   Metaphase I   Anaphase I   Telophase I

MEIOSIS II

Prophase II   Metaphase II   Anaphase II   Telophase II
(gametes)

Chromosomes still duplicated. Nuclear membrane fragments.   Chromosomes aligned at equator.   Chromatids divide. Chromosomes move towards poles.   Nuclear membrane reforms. Meiosis completed.

Delmar/Cengage Learning

FIGURE 1-5

Meiosis is the division of sex cells called gametes.

Organs and tissues combine to form systems in an animal's body. Many of these systems will be discussed further in following chapters. A partial listing of recognized systems follows:

1. The skeletal system consists of bone and cartilage tissue and serves as a support structure for the body.

2. The muscular system is made of muscle tissue attached to the skeletal system by tendons and allows the skeletal system to move.

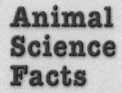

Animal Science Facts

Students often memorize the order of Linnaeus' classification hierarchy by reciting a sentence in which the first letter of each word aligns with the first letter of the words in the taxonomy. Following is one such example. Perhaps your instructor knows a different one. Can you think of your own original version?

King Philip Called Out for Giant Sandwiches

3. The respiratory and circulatory systems combine to ensure that internal cells get the oxygen needed to survive. These systems consist of the lungs and heart, respectively, along with miles of blood vessels that carry blood to the internal organs.

4. The digestive system consists of many organs including the stomach, intestines, and liver, which combine to break food into particles small enough to be carried by the blood stream and used by individual cells.

5. The nervous system is made up of the brain, spinal cord, and specialized nerve cells that carry electrical impulses to the brain for processing.

6. The endocrine system encompasses many specialized cells and glands that secrete substances, called hormones, into the blood and keep the body chemically balanced.

7. The reproductive system is closely tied to the endocrine system. The principal gland of the female system is the ovary, which produces ova (eggs) and secretes hormones that help to begin and sustain a pregnancy. The principal male gland is the testis, which produces sperm and hormones to ensure the proper development of sperms.

TAXONOMY

In the mid-1700s, Carolus **Linnaeus** developed a system for scientifically classifying all organisms according to their similarities. See figure 1-6. Since then, all organisms have been sorted by biologists into a structured classification or **taxonomy** system. In descending order, the seven steps of the system

FIGURE 1-6

Carolus Linnaeus developed a taxonomy for classifying living organisms.

Public Domain Painting by Hendrik Hollander, 1853

are: kingdom, phylum, class, order, family, genus, and species. The top rung of the taxonomy ladder consists of very broad categories called "kingdoms." There are five recognized kingdoms:

Monera (includes one-celled organisms)

Protista (includes slightly more advanced life forms, which can be single-celled or multicelled)

Fungi (includes molds, fungi, and lichens)

Plantae (includes all normally recognized plant forms)

Animalia (includes sponges, reptiles, insects, birds, mammals, and other animals)

The second rung of the ladder of classification is "phylum." Farm animals are included in the phylum *Chordata*, which includes all animals with a backbone.

The third rung is "class." Cattle, sheep, horses, and swine are in the class *Mammalia*, which also includes humans. Mammals are characterized by having hair- or fur-covered bodies and giving birth to young that are nourished by milk from their mothers.

Class *Mammalia* (mammal) is further divided into "orders" in the fourth rung of the classification ladder. Even-toed, hoofed animals, such as cattle, sheep, and swine, are in the order *Artiodactyla*. Horses are considered odd-toed, hoofed animals and fall under the order *Perissodactyla*. Dogs and cats are meat-eaters and are in the order *Carnivora*.

The last three rungs of the classification ladder are "family," "genus," and "species." These three classifications further sort animals into categories with similar characteristics.

Biologists commonly refer to organisms using the binomial system or **binomial nomenclature**. See figure 1-7. Binomial nomenclature, which was

COMMON NAME

	HOGS	CATTLE	SHEEP	HORSES
Kingdom	Animalia	Animalia	Animalia	Animalia
Phylum	Chordata	Chordata	Chordata	Chordata
Class	Mammalia	Mammalia	Mammalia	Mammalia
Order	Artiodactyla	Artiodactyla	Artiodactyla	Perissodactyla
Family	Suidae	Bovidae	Bovidae	Equidae
GenuS	Sus	*Bos*	*Ovis*	*Equus*
Species	*scrofa*	*taurus or indicus*	*aries*	*caballus*

FIGURE 1-7

Taxonomy of farm animals.

also proposed by Linnaeus, uses the organism's "genus" and "species" names written in Latin. For example, the binomial nomenclature name for swine is *Sus scrofa*. When using binomial nomenclature, the genus name is capitalized, but the species name is not. Both names are either underlined or italicized.

SWINE

Domestic swine (family *Suidae*), also known as *Sus scrofa* in binomial nomenclature, were domesticated from both the European wild boar and the East Indian pig. Pigs were first imported to the Western Hemisphere by Christopher Columbus. Imports of established European breeds were common during colonial times. Escaped pigs easily adapted to the wild, eating whatever they could find. Wild and more recently escaped pigs are still found in many parts of the United States. Animals that were once domestic and have returned to the wild are referred to as **feral**. Originally, domestic hogs were valued for their lard production, but since World War II, swine breeders have concentrated on selecting pigs for meat production. See figure 1-8.

FIGURE 1-8

Feral swine are found in parts of the United States.

FIGURE 1-9

Bos taurus cattle originated in Europe.

FIGURE 1-10

Bos indicus cattle originated in India and Africa.

CATTLE

Two species of cattle, *Bos taurus* and *Bos indicus*, make up the modern breeds of cattle. Normally, species are unable to interbreed and produce fertile offspring, but these two species of cattle are an exception. Both are from the family *Bovidae*. *Bos taurus*–derived cattle include short-eared cattle domesticated from the giant Aurochs, a primitive cattle of the European forests. See figure 1-9. *Bos indicus* cattle were first domesticated in India and Africa and are characterized by their long, drooping ears and hump above the shoulders. See figure 1-10. *Bos indicus* are also more tolerant of heat and some diseases than their *Bos taurus* counterparts. Early agriculturalists used cattle as a source of meat, milk, and draft power. The first cattle brought to the United States were of Spanish ancestry and easily adapted to the desert of the Southwest. Known as Longhorns, they could fend for themselves in the wild. Other cattle imported to the Colonial United States were more valued for their ability to pull a wagon or plow than for meat or milk. Later, European breeds specializing in meat or milk production were imported to improve the common stock.

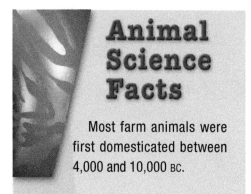

Animal Science Facts

Most farm animals were first domesticated between 4,000 and 10,000 BC.

SHEEP

Sheep (family *Bovidae*) were among the first livestock to be domesticated, probably because of their useful products, productivity, relatively small size, and noncompetitiveness with humans for food. Classified *Ovis aries*, there are over 200 distinct breeds of sheep scattered throughout the world. These breeds can all be traced to the wild sheep of Europe and Asia Minor. Spanish explorers imported the ancestors of domestic sheep to the Americas in the 1400s. See figure 1-11. Early on, producers concentrated their efforts on those sheep that produced the finest quality wool. This is due to the fact that wool is lightweight and imperishable over time, making it an ideal fiber source in undeveloped areas. Later, breeds were developed specifically for meat production.

HORSES

Hailing from the family *Equidae*, *Equus caballus*, or the domestic horse, first served humankind in Asia several thousand years ago. Nomads used the horse for milk, meat, transportation, and as a vehicle of war. Most likely, many types of wild horses were domesticated and mingled to form various breeds and types. Horses actually evolved in the United States but became extinct before the arrival of Columbus. Spanish explorers reintroduced horses to the Americans.

© Ewan Chesser. Used under license from Shutterstock.com

FIGURE 1-11

Domestic sheep originated from wild sheep of Europe and Asia Minor.

© Jeanne Hatch. Used under license from Shutterstock.com

FIGURE 1-12

Wild horses were once domesticated but now reproduce outside of captivity.

Escaped horses that turned wild (feral), were captured, and adopted by Native Americans. See figure 1-12. Later, large draft-type horses were imported from Europe to assist with farm labor. Horses were used for all types of transportation, including pulling barges, transporting freight, carrying mail, racing, and pleasure riding. Except for some societies, such as the Amish, who use them daily for labor and transportation, horses in the United States are now used mainly for pleasure.

SUMMARY

The basic unit of life is the cell, which acts as a little factory. Cells take up nutrients, excrete waste, do work, and reproduce. The process of normal cell division is mitosis, while sex cell division is called meiosis. Cells bond together to form tissues and organs, which in turn form systems in the body.

Organisms are classified into five different kingdoms, which are further divided into phyla, classes, orders, families, genera, and species. The final two classifications (genus and species) are used in binomial nomenclature to scientifically name organisms. Classification names should be italicized or underlined, and all should be capitalized except for the species name, which begins with a small case letter. A correctly written binomial name for the pig is *Sus scrofa*.

Most domesticated farm animals originated in Europe or Asia. The species have evolved to fit the needs of modern society. Formerly domesticated animals that have returned to the wild are called feral.

CHAPTER REVIEW

EXPERIENTIAL LEARNING OPPORTUNITIES

1. Job shadow a medical or veterinary technician who works with cells in a laboratory setting. Discover the significance of cells in the individual's work.

2. Research the domestication of a livestock specie and trace the history to modern-day use. Prepare a speech to present to the class or in a competition.

DEFINE ANY TEN KEY TERMS

apoptosis

binomial nomenclature

cell

cell membrane

centromere

chromosomes

cytoplasm

DNA

endoplasmic reticulum

feral

gametes

genes

golgi bodies

haploid

Linnaeus

lysosomes

meiosis

mitochondria

mitosis

nucleus

ribosomes

taxonomy

QUESTIONS AND PROBLEMS FOR DISCUSSION

1. List the four similarities of cells.

2. What part of the cell controls all cellular activity?

3. Genes are small pieces of the larger _____.

4. Chromosomes consist of strands of _____.

5. Name the two phases of the cell life cycle.

6. List the three phases of interphase.

7. What is the major difference between mitosis and meiosis?

8. The result of meiosis is four _____ cells instead of the two diploid cells, which result from mitosis.

9. Is cartilage a cell, tissue, or organ?

10. List five systems found in animals.

11. When did Linnaeus develop a system for classifying living organisms?

12. Write the seven classifications of Linnaeus' taxonomy in order, beginning with kingdom.

13. Name four farm animals that are mammals.

14. What farm animal listed in the chapter falls under the Perissodactyla order?

15. Correctly write the scientific name for swine.

16. Name one feral species found in the United States.

17. Contrast the two species of cattle.

18. Over _____ breeds of sheep are scattered throughout the world.

19. On which two continents were most livestock species domesticated?

20. Name the vocation highlighted in this chapter's career focus.

2

Biology of Growth and Development

GROWING UP BIG AND STRONG

Objectives

After completing this chapter, students should be able to:

▶ Draw and label a growth curve

▶ Compare and contrast bone with muscle growth

▶ Describe the process of adipose tissue (fat) deposition

▶ Differentiate between early- and late-maturing animals

▶ Discuss the economic importance of the ideal slaughter point

▶ Identify substances that increase the amount of muscle and decrease the amount of fat

Key Terms

adipocytes	conception	maturity
adipose tissue	estrogens	myoblasts
androgens	ewes	myofiber
barrows	gilts	myotube
beta agonist	glucose	presumptive myoblasts
boars	heifers	rams
bulls	intact	somatotropin
cartilage	intermuscular fat	subcutaneous fat
castrated	internal fat	steers
collagen	intramuscular fat	wethers

Career Focus

Ken was a communications major in college and always liked explaining his 4-H project animals to people when he exhibited market animals at the county fair as a high school student. After college, Ken landed a job as a communication specialist for the state Farm Bureau. When he arrived at work one day, a morning newspaper was already on his desk proclaiming in bold letters "Growth Hormone–Laced Beef Enters Food Supply." Ken knew he had a busy day ahead of him to project the facts about hormones and their relation to meat animal production.

INTRODUCTION

Many farm animals are raised solely for meat production. Therefore, producers want to raise animals that are heavily muscled without an excess of fat. In fact, market animal value and producer income are heavily determined by the amount of muscle and fat in the carcass. Factors that influence the amount of muscle and fat in the carcass include maturity pattern and sexual condition of the animal. The biology of muscle, fat, and bone growth, as well as natural substances that can alter muscle to fat ratio, is discussed in this chapter.

Delmar/Cengage Learning

FIGURE 2-1

Livestock come in many shapes and sizes.

THE GROWTH CURVE

Animals gain weight at different rates in the growth period from **conception** (time of fertilization) to **maturity** (adulthood). For example, a three-week-old pig may gain half a pound per day, while the same pig at 20 weeks of age could gain more than two pounds per day. To illustrate this concept, scientists often use the standard or S-shaped growth curve. See figure 2-2. Growth units, such as pounds, are represented on the vertical axis, and time units, such as days, weeks, or months, are represented on the horizontal axis. Early growth (region 1 of figure 2-2) before and shortly after birth is relatively slow. At a certain point after early growth, (point A of figure 2-2) body weight gain accelerates rapidly. This period compares to the normal teenage growth spurt and is shown in region 2 of figure 2-2. As animals reach maturity (point B of figure 2-2), body weight gain slows until a plateau (region 3 of figure 2-2) is reached at the animal's mature body weight.

Not all animals of the same species mature at the same body weight. Compare the growth curves of

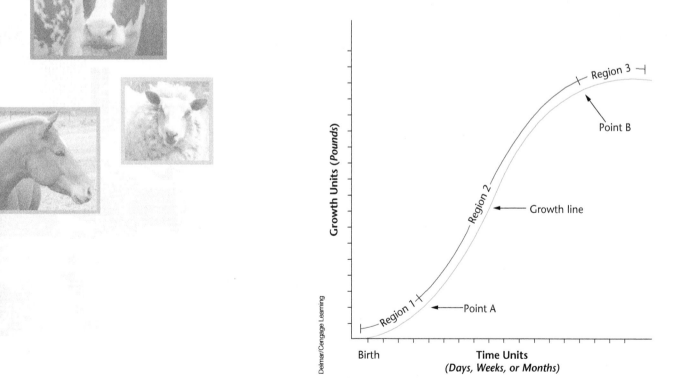

FIGURE 2-2
The standard S-shaped growth curve.

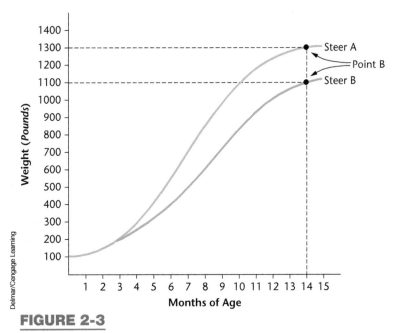

FIGURE 2-3

Growth curves of two steers with different mature body weights.

the two **steers** (male cattle with testicles removed) in figure 2-3. Notice that steer B's growth curve begins to plateau at a lower weight than that of Steer A (point B). If you draw a line from both B points back to the growth units line, you'll see that steer B has a lower mature body weight than steer A. Even though both animals reached mature body weight at the same time, steer B matured at a much lighter body weight.

Even with animals of the same mature body weight, the rate at which they reach that weight may vary. Compare two **barrows** (male swine with testicles removed) with different slopes to their growth lines. See figure 2-4. Again, concentrate on point B. At point B, the mature body weight is the same for both animals, but when you drop a line to the time units axis, you'll find that barrow A reached mature body weight in less time than barrow B. Therefore, barrow A grew faster.

Each animal has its own distinct growth curve. Some animals mature early on the time line, others late. Conversely, some animals mature at lighter weights than others. The slope of the growth line tells how fast the animal grew as it approached mature body weight. If you weigh an animal frequently from birth to maturity, you can plot a very similar curve for any normally growing animal.

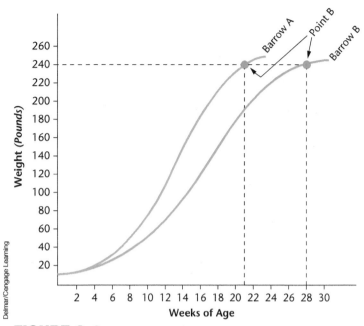

FIGURE 2-4

Growth curves of two barrows with the same mature body weight but different growth rates.

ECONOMICALLY IMPORTANT TISSUES

Muscle, bone, and fat are the three tissues found in greatest abundance in carcasses of slaughtered animals. These three tissues grow at different rates while the animal progresses from birth to maturity. The addition of muscle tissue makes up most of the carcass weight gain in young, rapidly growing animals. As that animal grows older and larger, the rate of muscle addition slows and the amount of fat added increases. At some point in the maturity process (if the animal is well fed), muscle growth slows drastically while fat development continues or even accelerates. This point is known as the ideal slaughter point. At the same time, bone growth maintains a fairly constant rate from birth to maturity.

As with total body growth, the rate of muscle and fat addition varies among individual animals. Deposition of fat, normally, occurs at a younger age for early-maturing animals than for later-maturing animals. Compare the muscle, fat, and bone growth curves in figure 2-5 and figure 2-6. Notice that the "ideal slaughter point" for the animal in figure 2-5 is at a heavier

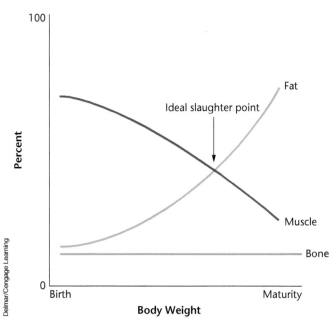

FIGURE 2-5

Percentage of muscle, bone, and fat in a large-framed, late-maturing animal.

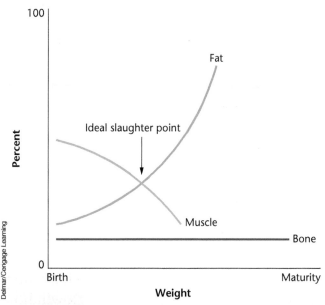

FIGURE 2-6

Percentage of muscle, bone, and fat in a small-framed, early-maturing animal.

weight and an older age than the animal in figure 2-6. Since animals convert feed more efficiently into muscle tissue than into fat, it is more economical for producers to raise animals that mature later and at heavier weights because feed costs per unit gain are lower. However, animals must have a minimal amount of fat to provide flavor and a satisfying eating experience to meat consumers.

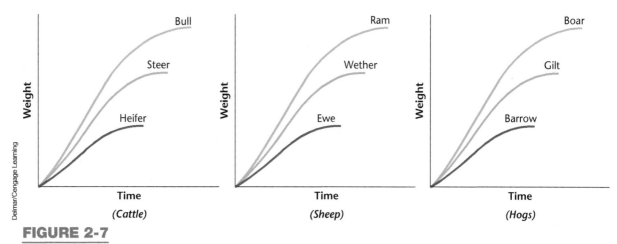

FIGURE 2-7

Maturity patterns among sexes in cattle, sheep, and swine.

The rate of muscle and fat addition varies depending on whether the animal is an **intact** (uncastrated) male, a female, or a **castrated** (testicles removed) male. See figure 2-7. Of these three sexual conditions, because of male hormones (**androgens**) that divert feed nutrients toward muscle development, intact males—**bulls**, **rams**, and **boars** (in cattle, sheep, and swine, respectively)—mature later and at heavier weights than either females or castrated males. The ranking of the latter two sexual conditions depends on the species of the animal. In cattle, steers (castrates) are later-maturing than **heifers** (young females). In sheep, **wethers** (castrates) are later-maturing than **ewes** (young females). In swine, **gilts** (young females) are later-maturing than barrows (castrates).

BIOLOGY OF MUSCLE DEVELOPMENT

There are three types of muscle: cardiac, smooth, and skeletal. Cardiac muscle is found only in the heart. Smooth muscle surrounds tubular organs, such as the intestines and esophagus. The most abundant and most economically important type of muscle in meat animals is skeletal muscle found attached to bones in the legs, back, shoulders, and elsewhere in the body. Skeletal muscle provides most of the meat produced for food.

The first cells that can be distinguished as premuscle cells in the embryo are called **presumptive myoblasts**. Presumptive myoblasts contain one nucleus and have the ability to divide. When presumptive myoblasts lose their ability to divide, they are called **myoblasts**. Toward the end of fetal development, many myoblasts fuse together to form **myotubes**. Myotubes look like mature muscle cells because they are long, narrow, and have many nuclei. Myotubes also contain certain proteins found only in mature skeletal muscle. Like myoblasts, myotubes do not have the ability to divide. Shortly before birth, myotubes begin functioning as mature muscle cells. See figure 2-8. Mature muscle cells are called **myofibers** and also have many nuclei. Myofibers are discussed further in Chapter 3. Myofibers do not have the ability to divide. The number of myofibers an animal is born with is the number it will have for the rest of its life.

Even though muscle cells cannot divide after an animal is born, muscle tissue continues to grow in width by the incorporation of satellite cells. Satellite cells are closely associated with the surface of the myofiber, but they retain their ability to divide. In that

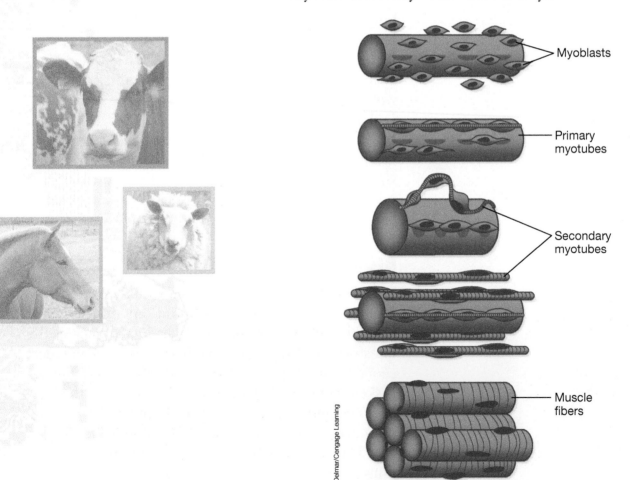

FIGURE 2-8

Development of muscle tissue before birth. Myoblasts, myotubes, and myofibers.

Animal Science Facts

The Ossabaw pig is a feral species native to the island of Ossabaw, off the coast of Georgia. Since food resources are limited on the island, the pigs adapt by gorging themselves during the short season when food is plentiful. As a result, they deposit a thick layer (measured in inches) of fat, which helps to sustain them when food supply is short. The Ossabaw pig has been useful to animal scientists researching adipose tissue development.

respect, satellite cells are very similar to presumptive myoblasts. However, satellite cells do not fuse to form myotubes. Instead, they are used by muscle tissue as a source of nuclei and associated DNA for increased muscle mass and accelerated muscle repair.

Muscle cells also grow lengthwise by stretch-induced muscle growth. As bones grow, muscles must lengthen to keep up, so muscle cells also lengthen.

BIOLOGY OF FAT DEVELOPMENT

Fat is commonly called **adipose tissue**, and individual fat cells are called **adipocytes**. Adipocytes are very large in diameter compared to other cells in the body, but the extra size consists almost entirely of stored energy. This energy is in the form of lipid—a thick, high-energy substance of about the same consistency as butter. Fat cells act as an extra gas tank for the animal. When the animal cannot consume enough energy to maintain the basic functions of life, the energy stored in adipocytes is released and used until more energy is consumed, or the energy demands of the animal are lessened. Alternately, when energy intake is higher than the animal needs, liver and fat cells store the extra energy as lipid in adipocytes.

Mature adipocytes do not divide, but pre-adipocyte cells do when necessary. As adipocytes store more lipids and reach maximum size, pre-adipocyte cells start dividing to produce a new crop of baby adipocytes to begin filling with lipid.

There are four major deposits of fat in the animal body. **Internal fat** within the body cavity surrounds the internal organs, kidneys, pelvic cavity, and the heart. Internal fat is the first to be stored as the animal matures and it acts to cushion organs from injury. See figure 2-9. **Intermuscular fat** is stored in the seams between muscles. See figure 2-10. **Subcutaneous** (under the skin) **fat** is stored between the hide and muscle and is usually called backfat. See figure 2-11. Most fat in pigs (80%) is stored under the skin while sheep and cattle have about equal amounts of subcutaneous and intermuscular fat volumes. Both intermuscular and backfat are stored about the same time, midway through the fat deposition process. The last fat deposit the animal develops is **intramuscular fat**.

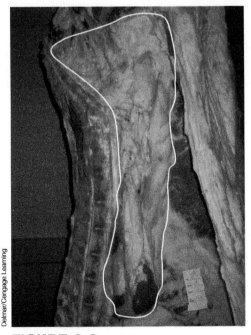

Delmar/Cengage Learning

FIGURE 2-9

Internal Kidney and pelvic fat in a beef carcass.

FIGURE 2-10

The large accumulations of fat in this photo are between muscle groups and are examples of intermuscular fat.

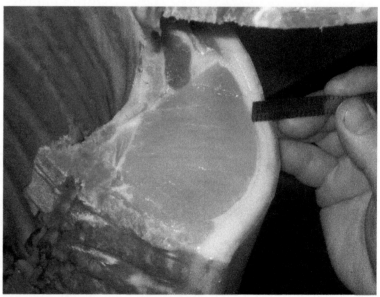

FIGURE 2-11

Measuring subcutaneous fat on a pork carcass.

See figure 2-12. It is more commonly called marbling and is found inside muscle bundles. Marbling accounts for the flavor and juiciness of meat, and thus highly marbled meat is desirable to consumers. The first three types of fat are undesirable in animals raised for meat production because of its low economic value relative to muscle. The order of fat deposition makes it difficult for producers to raise animals with highly marbled meat and small amounts of the other three types of fat. However, research suggests that animals can be genetically selected for small amounts of backfat and large amounts of marbling.

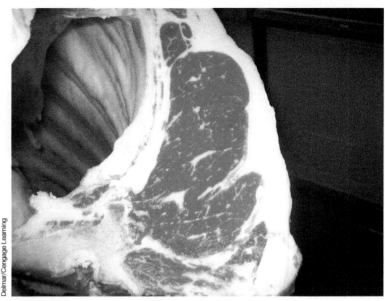

FIGURE 2-12

Flakes of intramuscular fat are called marbling and provide flavor to meat.

BIOLOGY OF BONE DEVELOPMENT

Bone growth in young animals is preceded by the growth of collagen strands. **Collagen** is a fibrous tissue composed of three strands of tropocollagen chemically bound together in a triple helix. Strand for strand, collagen is stronger than steel. Collagen is a major component of **cartilage**, the precursor to hard bones. Cartilage is formed at the end of long bones in a thin region called the epiphyseal growth plate. See figure 2-13. This growth plate contains many blood vessels that deliver raw materials to the rapidly dividing cartilage cells. As new cartilage is added within the growth plate older cartilage cells toward the interior of the bone die and are replaced by calcium. This process is called calcification. As the animal reaches maturity, cells in the growth plate stop dividing, the growth plate disappears, and the entire bone is calcified. Calcification of the growth plate signals the end of bone growth. At this point, the bone cannot grow any longer, but it may thicken as more calcium is added to the walls of the bone. Bone can also be remodeled throughout life with older bone getting degraded and being replaced by new bone.

At the time of birth, skeletons of baby animals are partially made of uncalcified cartilage. This provision

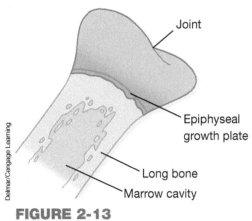

FIGURE 2-13

Long bones grow by cartilage formation and calcification at the epiphyseal growth plate.

by nature ensures that baby animals do not break any bones during the birth process. Shortly after birth, calcification begins in earnest.

CHANGING THE MUSCLE TO FAT RATIO

Skeletal muscle and fat are the most economically important tissues in the carcasses of animals raised for meat. Scientists have long searched for methods to produce animals with a higher proportion of muscle to fat. Before learning about such methods, we must first discuss theories on the distribution of nutrients.

For purposes of this discussion, think of all the protein in the animal's body as being present in one of two possible places: in muscle or in the blood stream. Simultaneously, protein is converted into muscle protein from blood protein and broken down from muscle protein into blood protein. The trick is to capture a higher percentage of total body protein in muscle rather than in the blood stream where it can be lost in the urine. In other words, if muscle manufacturing is greater than muscle break down, a net increase in muscle tissue will occur.

Likewise, **glucose** (a form of energy in the blood stream) is constantly converted into fat and broken down from fat. If the rate of fat production is greater than the rate of break down, an increase in adipose tissue will occur. Glucose is also required to build muscle tissue, so if more glucose is diverted to muscle accumulation there is less glucose available for fat formation.

Both these processes happen in unison. An animal storing muscle tissue will not store much fat and an animal storing a large amount of fat will not deposit much muscle tissue. Two substances have been shown to divert protein and energy toward muscle growth rather than fat accumulation.

SOMATOTROPIN

Somatotropin (ST), also known as growth hormone, is a naturally occurring protein hormone present in all animals, including humans. It is secreted by the pituitary gland, which is located at the base of the brain, and is species specific. This means that

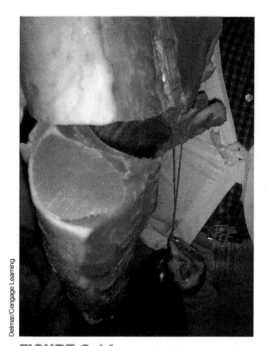

FIGURE 2-14

Growth hormone (somatotropin) is not legal to use in meat animals, but research has shown that it increases muscle volume and decreases fat accumulation.

somatotropin from one species (such as pigs) will not affect any other species (such as cattle, sheep, or humans).

Somatotropin regulates the growth of long bones, muscle, and fat. It is secreted in relatively large amounts in young, growing animals with the amount decreasing to a lower level at maturity. The greater the amount of naturally occurring somatotropin secreted during the growth phase, the larger and later-maturing the animal will be.

Somatotropin also works to make animals heavier muscled by increasing protein formation and decreasing protein breakdown, so a net gain in muscle mass results. See figure 2-14. Similarly, somatotropin works to make animals leaner by increasing the rate of fat breakdown so a net loss of fat mass results. In lactating dairy cattle, bovine somatotropin (bST) works to increase the amount of nutrients captured by the mammary gland (udder) for milk production.

Each animal has a naturally occurring amount of somatotropin in its bloodstream. If the amount is artificially increased, an animal that is genetically programmed to be fat, light muscled, and early-maturing can be changed into one that is leaner, heavier muscled, and later-maturing.

As natural somatotropin from dead animals cannot be collected and used on a large number of growing animals, scientists have genetically modified bacterial species to produce somatotropin in large amounts. This synthetic somatotropin is called recombinant somatotropin (rST) and has exactly the same effects as the somatotropin the animals produce themselves.

Somatotropin is a protein. If it were fed to animals, it would be digested like any other protein with no effect on the animal. Also, the effective life of somatotropin in the bloodstream is very short. Therefore, to have any measurable effect, it must be injected into the animal regularly. There is no difference in the safety or quality of meat or milk from animals injected with recombinant somatotropin when compared to meat and milk from animals that were not injected. In fact, it is impossible to tell the difference. Recombinant bovine somatotropin (rbST) is approved by the FDA for use in lactating dairy cattle. It is not legal for use in animals raised for meat.

Delmar/Cengage Learning

Animal Science Facts

Paylean®, a beta-agonist feed additive produced by Elanco is approved for market hogs that weigh over 150 pounds. Depending on the inclusion rate and length of time fed, pigs fed Paylean exhibit about 9 percent faster growth, an improvement of 14 percent in feed efficiency, 14 percent lower backfat thickness, and 11 percent more carcass lean tissue than pigs not fed the additive. For Paylean to have these desired effects, the protein content of the diet must be increased.

BETA AGONISTS

Beta agonists are synthetic substances similar to adrenaline (the hormone that courses through your body when you are suddenly frightened). As with somatotropin, they increase protein production into muscle and decrease protein breakdown from muscle. They also increase fat breakdown and decrease fat production. Beta agonists work by a different mechanism than somatotropin in the animal's body because the effects are additive when both are administered at the same time. Beta agonists are orally active and can be given to animals in the feed. The beta agonist ractopamine is approved for use on finishing swine and in finishing cattle.

SEX HORMONES (GROWTH PROMOTING IMPLANTS)

Sex hormones increase growth rate and lean tissue development while decreasing fat. Blood levels of sex hormones, such as androgens (in males) and **estrogens** (in females), can be increased through the use of implants in cattle and sheep. Small pellets containing the hormones are implanted under the skin behind the ear. See figure 2-15. Androgens are naturally secreted by the testes in the male, and estrogens are secreted by the ovaries in the female. When males are castrated, their natural supply of androgens ceases to exist. Implants are used by producers to restore some of the more desirable effects of androgens (such as leanness, increased growth rate, and less backfat) while avoiding some of the undesirable effects (such as unpredictable temperament and strong tasting meat). Implants also have similar effects on the carcasses of females even though the estrogen factories (ovaries) are still present.

Implants are legal for use in cattle and lambs fed for slaughter, although rectal prolapses are a common side effect in implanted lambs.

GENETIC SELECTION

The longest-used technique for improving the lean content in farm animals is through genetic selection. Animals that are lean and heavily muscled tend to produce similar offspring. As agriculture becomes more

Photo by Frank Flanders

FIGURE 2-15

Implants containing sex hormones are implanted behind the ear with this device.

competitive and economic circumstances urge producers to become more efficient, the search to find new ways to make animals produce more meat and milk and less fat will undoubtedly continue.

SUMMARY

Animals mature at different rates and at different body weights. Early-maturing animals that mature at lighter weights normally are fatter and lighter muscled than their later-maturing, counterparts. Growth curves can be developed for any growing animal and can tell us much about its maturity pattern. Muscle cells, fat cells, and bone tissues develop in distinctly different ways, but all are important components of the final carcass. Consumers demand that producers raise lean, meaty animals. The understanding of how muscle tissue and fat are proportioned in the body can assist the producer in raising animals with a higher percentage of lean muscle. Somatotropin, beta agonists, and sex hormones have been shown to improve the muscle to fat ratio.

CHAPTER REVIEW

EXPERIENTIAL LEARNING OPPORTUNITIES

1. Record weight gain of livestock projects. Create a growth curve for each. Draw conclusions based on gender. Pay particular attention to swine data.

2. Investigate concerns regarding milk labeling and use of rbST. Prepare an informative speech or deliver a point/counterpoint presentation for an issues forum.

DEFINE ANY TEN KEY TERMS

adipocytes	heifers
adipose tissue	intact
androgens	intermuscular fat
barrows	internal fat
beta agonist	intramuscular fat
boars	maturity
bulls	myoblasts
cartilage	myofiber
castrated	myotube
collagen	presumptive myoblasts
conception	rams
estrogens	somatotropin
ewes	steers
gilts	subcutaneous fat
glucose	wethers

QUESTIONS AND PROBLEMS FOR DISCUSSION

1. Draw a standard or S-shaped growth curve.

2. Animals of the same species always grow to maturity at the same rate. True/False.

3. Name the three economically important tissues found in an animal carcass.

4. When does the ideal slaughter point occur in animals?

5. Which animal tissue continues to grow at a steady rate from birth through maturity?

6. Why is it more economical for producers to raise animals that mature later and at heavier weights?

7. Which tissue is the precursor to hard bone?

8. Which mature later, intact or castrated males?

9. In which of the three species, swine, cattle, or sheep, do females mature later than castrated males?

10. Name the three types of muscle tissue.

11. List the four areas of fat deposition.

12. Explain the calcification process of long bone growth.

13. _____ is a form of energy in the bloodstream.

14. What is the only FDA-approved farm-animal use of somatotropin?

15. Somatotropin is secreted by the _____ gland.

16. Recombinant somatotropin is genetically engineered. True/False.

17. Is somatotropin a protein or a fat?

18. When beta agonists are given to animals, protein manufacture increases/decreases and fat manufacture increases/decreases. Write correct answers.

19. Ractopamine is approved for _____ swine and cattle.

20. Implants are placed in which part of the animal's body?

Muscle and Meat Biology

BULKING UP

Objectives

After completing this chapter, students should be able to:

▶ Describe the difference between hypertrophy and hyperplasia

▶ Draw and label the muscle cell

▶ Describe the functions of a muscle cell

▶ List and distinguish between the types of muscle cells

▶ Explain postmortem changes in muscle

▶ Discuss two factors that can cause differences in meat quality

▶ Describe two processes that improve meat tenderness

Key Terms

actin	hyperplasia	postnatal
aerobic exercise	hypertrophy	rigor mortis
anaerobic exercise	muscle fatigue	sarcomere
endomysium	myosin	sarcoplasmic reticulum
epimysium	perimysium	titin
glucose	porcine stress syndrome	tropomyosin
glycogen	postmortem	troponin

Career Focus

Frank is a gym rat who spends two hours lifting weights after school each day. He has increased his strength tremendously in the past year and has built well-developed biceps. Yet he wonders how this occurs and why some of his friends, who spend just as much time training, have not seen such remarkable changes in their physiques. One day in his FFA Animal Science class, his instructor shows a picture of a Belgian Blue bull. The muscles of this animal are phenomenal! Frank decides to study muscle biology to understand why some people and animals have incredible muscle mass while others do not. After an undergraduate career in Animal Science, Frank pursued a PhD in muscle physiology and now heads a laboratory team studying ways to reduce the rate of muscle degeneration in the elderly.

INTRODUCTION

The production of meat for consumption is the main reason for humans to raise many livestock species. Meat is simply dead muscle tissue. Muscle tissue is approximately 72 percent water, 20 percent protein, 7 percent fat, and 1 percent minerals. To better understand meat and meat products, a basic knowledge of muscle biology is critical.

© Margo Harrison. Used under license from Shutterstock.com

FIGURE 3-1

Muscle development is important for all livestock, not just meat producing animals.

FIGURE 3-2
Exercise helps muscles to grow and develop.

HYPERPLASIA AND HYPERTROPHY

You may be familiar with names of whole muscles like the biceps in the upper part of your arm. Did you ever wonder how that muscle got there, and how it grew at the same rate as the rest of your body? Individual muscle cells differentiate and divide before birth in a process caller **hyperplasia**. The hyperplastic process is complete before birth, so an animal is born with all the muscle fibers it will ever possess. However, after birth, as an animal's skeleton lengthens, muscles must lengthen to keep up. Muscles also get larger in diameter as animals mature. Both types of **postnatal** (happening after birth) muscle growth are called **hypertrophy** (enlargement). Increase in bone length actually causes muscle cells to lengthen at each end. This type of hypertrophy is called stretch-induced hypertrophy. Muscle cells (and thus whole muscles) increase in width through exercise-induced hypertrophy. See figure 3-2.

MUSCLE AND MUSCLE CELLS

A single muscle cell is called a myofiber. Each myofiber is individually wrapped by a thin film of connective tissue called the **endomysium**. Bundles of myofibers are wrapped together by a connective tissue sheath called **perimysium**. These bundles of myofibers make up whole muscles that are wrapped by a connective tissue sheath called **epimysium**. All the connective tissues surrounding myofibers, bundles of myofibers, and whole muscles consist of a protein called collagen. See figure 3-3.

Each myofiber is divided into striped units called **sarcomeres**. The sarcomere is a tiny segment of the myofiber that shortens when muscles contract and elongates when muscles relax. Z-lines define the beginning and end of a single sarcomere as shown in figure 3-4. Sarcomeres are predominantly made of five muscle proteins: **myosin**, **actin**, **troponin**, **tropomyosin**, and **titin**. The thicker filament in figure 3-4 represents myosin and is referred to as the A band of the sarcomere. The thin

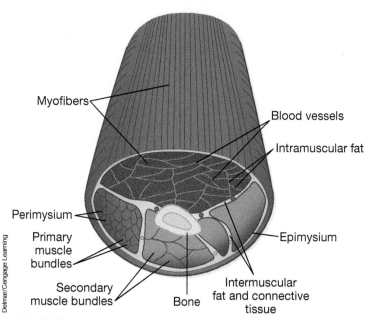

FIGURE 3-3

Arrangement of myofibers, muscle bundles, and connective tissue within a whole muscle.

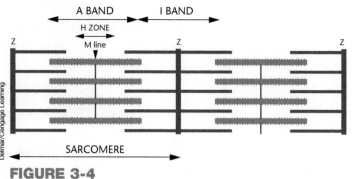

FIGURE 3-4

A sarcomere is the unit of a muscle fiber that contracts.

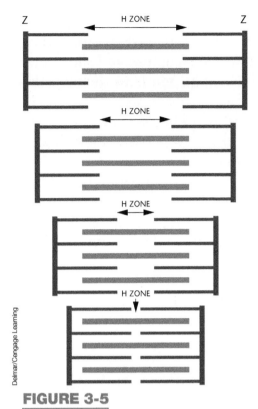

FIGURE 3-5

The sarcomere changes shape during contraction.

filament in figure 3-4 represents actin and is referred to as the I band of the sarcomere. While troponin and tropomyosin help regulate muscle contraction, titin appears to prevent overstretching of the muscle.

Of these five proteins, the thick and thin filaments (myosin and actin, respectively) are the most important to muscle contraction. As the sarcomere contracts, myosin stays stationary and attachments located at the ends of the myosin molecule, called "myosin feet," pull the actin filaments closer together, shortening the area called the H zone and causing the sarcomere to shorten. See figure 3-5. Thousands of sarcomeres shortening in unison cause the entire muscle to contract and shorten. When the muscle relaxes, each sarcomere resumes its original position and the muscle resumes its normal shape.

Calcium allows muscle contraction to occur. Each myofiber is surrounded by a reservoir of calcium called the **sarcoplasmic reticulum** (SR). When the muscle receives a message to contract from a nerve, the SR releases calcium into the myofiber. This causes the myosin feet to pull the actin filaments closer together. When the nervous impulse ends, calcium flows back into the SR and the muscle relaxes. The relaxation process requires no energy; actin and myosin molecules simply slide back to their original positions.

The whole process of contraction is fueled by either **glucose** or **glycogen**. Glucose is blood sugar, which can be obtained directly from the blood stream. Glycogen, or stored glucose, is found within the muscle. If the muscle runs out of stored glycogen or cannot get enough glucose from the bloodstream, it goes into a state of exhaustion called **muscle fatigue**.

TYPES OF MUSCLE CELLS

There are two main types of myofibers, distinguished by color and relative speed of contraction. The first type is called fast, white fibers. See figure 3-6. These fibers are large in diameter and have few blood vessels running through them. Fast, white fibers use stored glycogen within the muscle cell as their primary energy source. Large muscle groups consist of mostly fast, white fibers. The breast meat of poultry is composed almost entirely of fast, white fibers. **Anaerobic exercise**, which is intense and short term (for example, weight lifting), tends to develop this type of muscle fiber.

The second type of myofiber is slow, red fibers. See figure 3-7. These fibers get most of their energy directly from glucose in the blood stream. Therefore, these fibers have many blood vessels running through them and are red in appearance. Poultry drumsticks are composed mostly of slow, red fibers. **Aerobic exercise**, which is long term and less intense (for

FIGURE 3-6

Chicken breasts are composed mostly of white muscle fibers.

FIGURE 3-7

Chicken legs contain predominantly red muscle fibers.

example, jogging), develops this type of muscle fiber. To visualize the difference between fiber types, compare the color of a drumstick (slow, red fibers) to breast meat (fast, white fibers) from a Thanksgiving turkey.

The type of myofiber that predominates in an animal is largely controlled by genetics and varies by muscle group. For example, muscle fibers present in the loin (back muscle) tend to be fast and white, while muscle fibers in the legs tend to be slow and red. Moreover, heavily muscled animals have more fast, white fibers, which are larger in diameter and take up more space. Several intermediate types of myofibers also have properties of both fast, white and slow, red fibers. If an animal is subjected to stresses that favor a specific fiber type, fibers have the ability to slowly transform through the intermediate types toward the favored type. For example, wild sheep have mostly slow, red fibers because their existence demands they run away from predators. If the same animals were placed in captivity where the need to run away was greatly diminished, some of the slow, red fibers would begin to remodel and change to fast, white fibers.

MUSCLE TO MEAT

When an animal is slaughtered, blood stops flowing to muscle tissue. Thus, the immediate energy supply of glucose is cut off. Muscle cells continue to attempt contraction using glycogen as an energy source. This process produces lactic acid as a waste product. Lactic acid is normally carried away by the blood stream, but since that avenue is no longer available, the acid builds up in the muscle tissue causing the pH of the muscle to decline. Normal pH in a myofiber is 7.0. See figure 3-8. When the pH drops to around 5.5 due to the accumulation of lactic acid, muscles cease to use glycogen. Furthermore, the lactic acid build-up prohibits sarcomeres from moving and the muscle becomes stiff. This state of **postmortem** (after death) muscle stiffness is called **rigor mortis**.

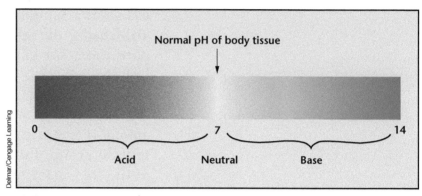

FIGURE 3-8

The pH scale. Numbers greater than seven indicate basic substances. Numbers less than seven indicate acidic substances.

DIFFERENCES IN MEAT QUALITY

The treatment of animals before and after slaughter can affect the final quality of meat. If muscles undergo strenuous exercise immediately before slaughter, muscle glycogen reserves are depleted. Therefore, there is not enough glycogen present in the muscle after slaughter to reduce the pH to normal postmortem levels. Meat from such animals exhibits a dark, firm, and dry appearance and is known as DFD. See figure 3-9C.

On the other hand, if muscle pH drops too rapidly after slaughter, meat develops a pale, soft, and watery appearance. Pale, soft, and exudative or watery meat (PSE) is often associated with pigs that have a condition known as **porcine stress syndrome** or PSS. Compare PSE ham in figure 3-9A with a normal ham

FIGURE 3-9A

Pale, soft, exudative (PSE) ham.

FIGURE 3-9B

Normal quality ham.

FIGURE 3-9C

Dark, firm, dry (DFD) ham.

Animal Science Facts

The average American meat consumption is the highest in the world.

Average American meat consumption, 2007

Chicken: 86.5 pounds

Beef and veal: 65.5 pounds

Pork: 50.5 pounds

Turkey: 17.3 pounds

Lamb and mutton: 1.1 pounds

Source: U.S. Department of Agriculture

in figure 3-9B. Pigs with this condition have a genetic abnormality in which they have leaky calcium channels between the SR and myofiber. In other words, calcium can flow from the SR into muscle causing a contraction without a nervous impulse. After slaughter, glycogen is converted to lactic acid more rapidly than normal, causing a quick drop in pH and ultimately, PSE meat. Both DFD meat and PSE meat are very unappealing to consumers and undesirable to packers.

MEAT QUALITY IMPROVEMENT

Muscles are like ropes. If all the fibers of the rope are continuous, the rope is very strong. Likewise, if all the myofibers and sarcomeres of a muscle are intact, the muscle is very strong—or tough to chew. However, there are several ways to break the myofiber and sarcomere structure after slaughter and, in the process, make meat more tender.

Meat becomes more tender with age. In fine restaurants, you will often see "aged beef" on the menu. Culinary experts identify this to be "tender beef." The reason aging makes meat more tender is because of enzymes that work by breaking some of the Z-lines between sarcomeres. The longer the meat is aged (held in a cooler after slaughter for up to 21 days), the more the Z-lines are broken by the enzymes, and the more tender the meat. See figure 3-10.

Freezing is a second way to tenderize meat. Remember that muscle is 72 percent water. When meat is frozen, the water present in the meat also freezes. When water freezes it expands, breaking apart sarcomeres and making the meat more tender when it is thawed and cooked.

In some large beef packing plants, an electrical current is passed through the carcass immediately after slaughter. The current causes all the muscles in the carcass to contract sharply and increases the rate of glycogen depletion. The pH of stimulated carcasses drops faster than normal, but in beef, this generally improves the color of the lean meat. Electrical stimulation also breaks apart muscle cell structure, improving tenderness.

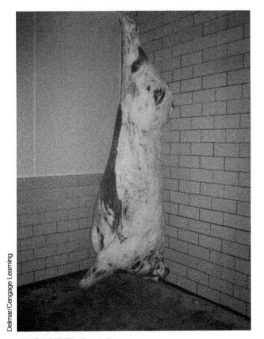

Delmar/Cengage Learning

FIGURE 3-10

Beef tenderness improves with increased time in the cooler.

SUMMARY

Muscle cells differentiate and divide by hyperplasia before birth, and muscle tissue enlarges after birth through hypertrophy. Individual muscle cells are called myofibers. Myofibers are divided into striped units called sarcomeres, which shorten when muscles contract. Actin and myosin are the two most important proteins found in the sarcomere. They cause muscle contractions. Contractions of muscles use either stored energy in the form of glycogen or energy from the bloodstream in the form of glucose. Anaerobic muscle fibers use mostly stored glycogen and are called fast, white fibers. Aerobic muscle fibers use mostly blood glucose and are called slow, red fibers.

After slaughter, muscle continues to mobilize immediately available energy reserves. This produces lactic acid and causes the pH of the meat to fall. Preslaughter animal handling, genetics, and the amount of energy available to muscle cells after death can alter the quality of the final meat product. Dark, firm, and dry meat and pale, soft, and watery pork are examples of muscle quality problems. Aging, freezing, and electrical stimulation are examples of ways to improve meat quality.

CHAPTER REVIEW

EXPERIENTIAL LEARNING OPPORTUNITIES

1. Get permission to tour a slaughter plant.

2. Design a research project that compares and contrasts various methods of meat tenderization.

DEFINE ANY TEN KEY TERMS

actin	perimysium
aerobic exercise	porcine stress syndrome
anaerobic exercise	postmortem
endomysium	postnatal
epimysium	rigor mortis
glucose	sarcomere
glycogen	sarcoplasmic reticulum
hyperplasia	titin
hypertrophy	tropomyosin
muscle fatigue	troponin
myosin	

QUESTIONS AND PROBLEMS FOR DISCUSSION

1. A single muscle cell is a _____.

2. List, in order, the connective tissues that surround myofibers, bundles of myofibers, and whole muscle.

3. _____ shorten when muscles contract and elongate when muscles relax.

4. Which two muscle proteins help regulate contraction?

5. Calcium, which causes muscle contraction, is stored in the _____.

6. _____ is blood sugar, while _____ is reserve sugar that is stored in the muscle.

7. Muscle fatigue occurs when muscles run out of _____ and can't get enough glucose from the blood stream.

8. Name the type of myofiber that a weight lifter would develop.

9. Aerobic muscle fibers use _____ for energy.

10. Anaerobic muscle fibers use _____ for energy.

11. Give an example of poultry meat that contains mostly slow, red myofibers.

12. Normal muscle pH is _____.

13. Explain why rigor mortis occurs.

14. What does DFD refer to in terms of meat appearance?

15. Describe the appearance of meat from a hog that has porcine stress syndrome.

16. What intramuscular occurrence makes meat more tender?

17. The text describes several methods used to tenderize meat. Name two.

18. True or False. Meat becomes more tender with age.

19. Where is meat aged?

20. What does electric current do to a beef carcass?

Biology of Digestion

I CAN'T BELIEVE I ATE THE WHOLE THING!

Objectives

After completing this chapter, students should be able to:

▶ List the six essential nutrients and their functions within the body

▶ Describe the process of digestion

▶ Compare and contrast digestion in pigs, poultry, cattle, sheep, and horses

▶ Explain the relationship between gross energy, digestible energy, metabolizable energy, and net energy

▶ Compare and contrast typical feedstuffs fed to ruminants and nonruminants

Key Terms

abomasum	macrominerals	reticulorumen
amino acid	microminerals	ruminant
cecum	monogastric	silage
chyme	net energy	simple carbohydrates
complex carbohydrates	omasum	symbiotic
crop	peptidases	villi
gizzard	peristalsis	volatile fatty acids
hydrochloric acid	proventriculus	

Career Focus

Megan's agriculture teacher, in a lecture about nutrition, insisted that water is the most important nutrient in the body. That comment got Megan thinking about the amount and quality of water that her show goats were drinking. Were they drinking enough? Was there anything in the water that kept her goats from growing as fast as they should? Megan decided to compare the amount of water her goats consumed to that consumed by her friends' goats of approximately the same size. To do this, she developed a protocol for watering each day. She and her friends carefully measured out exactly one gallon of water in the morning and again at night, then recorded how much was left before the next watering. When the numbers were tabulated for two weeks, Megan discovered that her goats drank about 25 percent less water than her friends' goats. To investigate why, Megan collected some water from her barn and sent it for chemical analysis. The results showed that the water was fine. Megan was stumped. As a last resort, she called her county extension educator, Sally, to help solve the problem. The next day Sally came to the farm, took one look at Megan's goat watering system, and identified the problem immediately. Megan's water tub was very dirty. Sally asked Megan what career she was interested in, and Megan answered that she wanted to do something with agriculture and wanted to help people. Sally suggested that Megan become an extension educator like herself. Megan is exploring the schooling required to pursue that line of work.

FIGURE 4-1

Forages are the backbone of ruminant nutrition.

INTRODUCTION

Animals use feed to produce products (meat, wool, milk, or eggs) and energy for work. The process of digestion reduces feed particle size and simplifies chemical composition of the feed for absorption by the animal. After absorption, the bloodstream transports the digested nutrients to body cells. There they are used to sustain life or are converted into animal products. Cattle and sheep have a special stomach chamber called a *rumen*. It contains multitudes of bacteria that digest some feedstuffs that other animals, such as swine and poultry, cannot digest. Horses also have a population of microbes to assist in the digestion of such feedstuffs.

ESSENTIAL NUTRIENTS

Six essential nutrients must be present in animal diets to sustain life: water, protein, vitamins, minerals, carbohydrates, and fats. Some must be ingested in large amounts, some in small amounts, but all are critical. Inadequate amounts of any of these essential nutrients will result in less-than-optimal growth and production.

WATER

Water is the most abundant nutrient in the body. Between 40 and 80 percent of total body weight is comprised of water. It is the medium in which all chemical reactions occur within the cells of the body. Water is also a major component in many body fluids, including blood and the lubrication fluid that surrounds bone joints. Water is lost through urine, as water vapor while breathing, as feces, and in some cases, through perspiration. Animals freely consume enough water to support bodily functions if a clean, fresh supply is available at all times. See figure 4-2. In the Career Focus section, Sally easily identified that Megan's water tub hadn't been cleaned for some time. The crusted manure and filth undoubtedly kept Megan's goats from drinking as much as they should. Megan immediately noticed an increase in feed intake and growth rate as soon as she cleaned the watering system.

PROTEIN

Protein promotes the growth and repair of body cells. Protein consists of strings of building blocks called **amino acids**. Nitrogen is the key element in amino acids. With the exception of animals that have a special stomach chamber called a rumen, amino acids cannot be manufactured within the body but must be consumed in the feed. Some amino acids critical to bodily functions are called essential amino acids. A deficiency in one or more essential amino acids can cause reduced growth rate or, in extreme cases, death.

© Michael Ledray. Used under license from Shutterstock.com

FIGURE 4-2

Animals need access to plenty of clean water at all times.

FIGURE 4-3

Clover is a good source of protein for ruminants.

Common plant feeds that contain high levels of protein include soybean meal, young alfalfa hay, and distillers grains (by-products from distilleries) See figure 4-3. Most feeds derived from animal products are also high in protein. Animal protein feeds include fish meal, meat meal, and milk products.

Bacteria living within a unique stomach chamber called the rumen in cattle and sheep (**ruminants**) have the ability to manufacture new proteins if fed nitrogen and an energy source. This process will be discussed in further detail later in the chapter.

During digestion, feed proteins (from plant, animal, or bacterial sources) are disassembled into individual amino acids. Amino acids are transported across the intestinal wall and are carried by the blood stream to body cells where they are reassembled into animal proteins, such as muscle, hair, or collagen. Milk and eggs also contain proteins manufactured by animal cells. In fact, a major job of nearly every body cell is to manufacture some type of protein. In the event of reduced feed energy intake, animals have the ability to convert muscle protein to an energy source for body cells.

VITAMINS

The two classes of vitamins are fat soluble and water soluble. Fat-soluble vitamins include vitamins A, D, E, and K. Vitamin A is especially important for eyesight and maintenance of skin cells. Vitamin D is essential for bone and tooth development. Vitamin E is important to red blood cell structure and triggers the energy metabolism of other cells. Vitamin K is an essential blood clotting factor. Fat-soluble vitamins enter animals through fats they consume and are stored in adipocytes. These vitamins are found in plant-based feeds in a preliminary or precursor form. Animals have the ability to convert this precursor form to a final, useable vitamin.

Water-soluble vitamins include biotin, choline, folacin, inositol, niacin, (the B-vitamin family), and vitamin C. The water-soluble vitamins assist in many bodily functions. Some can be made available within the animal's body from food-borne precursors. Others are manufactured by bacteria within the rumen. Water-soluble vitamins cannot be stored within the body and must be consumed or manufactured on a regular basis.

Vitamin needs are normally met in farm livestock by feeding vitamin-mineral premixes either free-choice (mineral mix consumed at animal's discretion) or mixed with other feeds.

MINERALS

There are two groups of minerals. Major minerals (**macrominerals**) are needed in the diet in relatively large amounts (grams per day). Trace minerals (**microminerals**) are required in smaller amounts (thousandths or millionths of a gram per day).

The macrominerals include salt (a mixture of sodium and chlorine), calcium, phosphorus, magnesium, potassium, and sulfur. Salt and potassium are involved in maintaining fluid balance inside and outside of cells. Calcium, phosphorus, and magnesium are important to bone structure and development. Sulfur is an important element in some amino acids.

Microminerals include chromium, cobalt, copper, fluorine, iodine, iron, manganese, molybdenum, selenium, silicon, and zinc. The trace amounts of these minerals are often present in normally fed grains and forages, but producers routinely supplement them in the diet to be sure that enough are consumed. See figure 4-4.

Photo courtesy of Penn State College of Agricultural Sciences

FIGURE 4-4

Beef cattle and sheep often consume commercially prepared minerals free-choice.

FIGURE 4-5

Grains provide easily digestible carbohydrates to both ruminants and nonruminants.

FIGURE 4-6

Hay contains cellulose and hemicellulose that cannot be digested by nonruminants.

CARBOHYDRATES

The major element in carbohydrates is carbon. **Simple carbohydrates**, such as starches and sugars, are used as a quick energy source by the animal. See figure 4-5. Simple carbohydrates are made up of short chains of several glucose molecules chemically bound together. The digestion process quickly disassembles these glucose chains to single molecules of glucose. Glucose is then absorbed by the blood stream, which carries it to body cells. Glucose is used as fuel for chemical reactions that produce animal products, such as milk, muscle, or eggs, or energy for work.

Two **complex carbohydrates**, cellulose and hemicellulose, cannot be digested by simple-stomached animals, such as swine and chickens. See figure 4-6. However horses and ruminants possess specialized digestive structures that house bacterial species capable of disassembling complex carbohydrates into simpler, digestible compounds.

FATS

Fats are a concentrated energy source that can be fed to animals in limited amounts. See figure 4-7. In

FIGURE 4-7

Vegetable oil or rendered animal fat can be fed in small amounts to increase the energy density of the diet.

fact, fat contains 2.25 times as much energy as carbohydrates or proteins. Fats are used in animal diets to boost energy levels derived from feed without increasing the volume. They are also important carriers of fat-soluble vitamins.

ENERGY METABOLISM

When animals consume energy from a feedstuff, not all the energy is available for growth and production. Energy is measured in kilocalories, that is, the amount of energy required to heat one liter of water by 1 degree C. If we simply burn the feedstuff and measure the amount of energy it contains, that amount is known as the gross energy (GE). However, some energy is not utilized by the animal and passes through the digestive tract. The amount of energy remaining after digestion, subtracted from GE gives the quantity of energy that was utilized by the animal. This retained energy is called digestible energy (DE). To put this concept in mathematical form: GE − Undigested energy from indigestible feedstuffs = DE.

Furthermore, not all digestible energy can be used for weight gain or production. A certain amount of energy is lost in urine and as gasses produced during the digestion process. This amount of energy can be measured by scientists and is subtracted from DE to arrive at metabolizable energy (ME). This portion of the equation can be summarized by the formula: DE − urinary energy − gaseous energy = ME. When we subtract the heat produced during the digestion process from ME, we arrive at the amount of energy left for the animal's cells to use. The energy available for animal use is called **net energy** (NE), so ME − heat energy = NE.

Some of the NE available to the animal is used for cell metabolism, for voluntary activity, and to keep the animal warm or cool. This portion of NE is called NE-maintenance. The remaining net energy is available for growth or production of animal products. This is called NE-gain or NE-lactation for dairy cows. See figure 4-8.

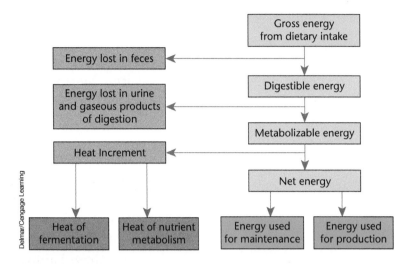

FIGURE 4-8

Only a portion of the energy an animal consumes is available for growth and production.

THE DIGESTION PROCESS

Different parts of the digestive tract and accessory organs serve different purposes during the digestion process. Particle reduction begins in the mouth. Teeth physically break apart food particles while enzymes in saliva break down simple carbohydrates. Food that is swallowed then moves down the esophagus to the stomach. The esophagus is simply a tube made of smooth muscle linking the mouth with the stomach. Food moves through the esophagus by **peristalsis**— involuntary smooth muscle contractions.

When food particles get to the stomach, they are mixed with **hydrochloric acid**—an extremely acidic substance that begins to break down proteins into shorter chains of amino acids. The low pH of the stomach is caused by hydrochloric acid. The acid kills most bacteria that are ingested with the feed and prevents them from infecting the lower portions of the digestive tract.

Partially digested material, called **chyme**, then moves into the small intestine. The small intestine is divided into three parts: duodenum, jejunum, and ileum. Much of the remaining digestive activity occurs in the duodenum, while food is absorbed into the bloodstream from the jejunum and ileum. Enzymes produced by the pancreas, called **peptidases**, are mixed with the chyme in the duodenum. Peptidases further reduce the amino acid chains that escape from the stomach. The reduced amino acids can then be absorbed by the many small blood vessels found in the walls of the intestine. Amylases are also secreted by the pancreas to further break down carbohydrates into simple glucose. Lipases secreted by the pancreas, and bile secreted by the liver and stored in the gall bladder, break up fat molecules into a form that can be absorbed. Pancreatic enzymes and bile are both secreted into the small intestine through ducts connecting to the duodenum.

The absorption process is aided by the structure of the small intestine. The interior of the small intestine contains many folds, which increase the surface area, thereby exposing more blood vessels to absorb digested food particles. Microscopic finger-like projections on these folds, called **villi**, further increase the surface area for absorption. See figure 4-9.

Indigestible components of the diet, as well as water, enter the large intestine or colon from the small intestine. At the beginning of the colon is the organ called the **cecum**. In humans, the cecum is known as the appendix. The cecum has no known function in cattle, sheep, or hogs; but it supports a population of digestive bacteria in horses. In the colon, water is absorbed by the blood stream, and undigested material is packaged for excretion.

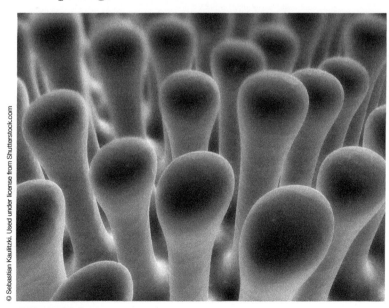

© Sebastian Kaulitzki. Used under license from Shutterstock.com

FIGURE 4-9

Villi increase the surface area of the digestive tract and improve the rate of nutrient absorption.

TYPES OF DIGESTIVE SYSTEMS

The basic digestive tract of all species is the one described in the last section. However, most species of farm animals have slight modifications to this theme.

SWINE

Swine are **monogastric** (one- or simple-stomached) animals that have digestive systems most similar to the one just described. See figure 4-10. The digestive tracts of pigs are designed to use high-energy feeds. Therefore, diets for pigs are high in carbohydrates, for example, feed grains. Diets containing large amounts of forages, for example hay, are not particularly suitable for monogastric animals like pigs. The human and swine digestive systems are remarkably similar. Typical swine diets include ingredients like shelled corn, milo, soybean meal, vitamins, and minerals.

POULTRY

Poultry are also monogastric. However, their digestive tract is different from that of a pig or human. See figure 4-11. The upper portion of the digestive tract of poultry includes two additional structures: the **crop** and **gizzard**. The crop, located at the base of the neck, serves as a storage area for recently ingested feed. From the crop, food passes to the **proventriculus**, or true stomach. The proventriculus serves the same

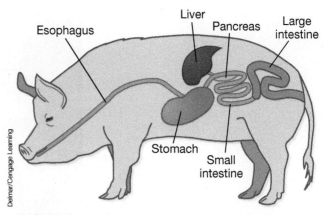

FIGURE 4-10

The swine digestive tract is very similar to that of a human.

functions of acid-mixing as the true stomach in other monogastric animals. Food particles pass from the proventriculus to the gizzard. Since poultry have no teeth, the gizzard is used to grind coarse feed particles. Since both are monogastrics, feedstuffs used for poultry are similar to those used in swine diets.

CATTLE, SHEEP, AND GOATS (RUMINANTS)

The digestive systems of cattle, sheep, and goats (ruminants) differ from monogastrics. See figure 4-12. Ruminants have adapted to diets containing large

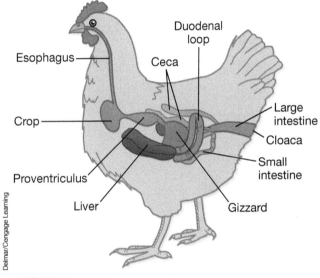

FIGURE 4-11

Poultry are monogastric animals, but the digestive anatomy is slightly different from that of a pig.

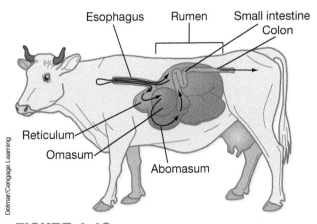

FIGURE 4-12

The ruminant digestive system is designed to digest large volumes of forages.

amounts of forages. Forages are plant-based feeds, which include stems and leaves. Some examples of forages are hay, pasture grasses, legumes, and **silage** (fermented forage). The stems and leaves of these feeds contain sources of complex carbohydrates, cellulose, and hemicellulose, which monogastrics cannot digest. However, ruminants can digest these complex carbohydrates with the aid of a remarkable stomach chamber called the rumen and houseguest bacteria present in the rumen.

The rumen is the first part of a four-compartment stomach that is located in the digestive tract. It precedes the small intestine. In cattle, the rumen is by far the largest of the four compartments, holding up to 50 gallons. The first two stomach parts, the rumen and reticulum are often referred to together as the **reticulorumen**. The reticulum is a small pouch located on the side of the rumen. It functions as a trap for foreign materials (nails, glass, or wire) which could be mistakenly eaten by the animal. The third compartment of the ruminant stomach is referred to as the **omasum** or manyplies. The omasum absorbs water from the chyme. The fourth compartment, the true stomach or **abomasum**, contains high levels of acid similar to that of monogastrics.

Ruminants are equipped with a dental pad in lieu of top incisor teeth. Ruminants use this adaptation to tear off large bites of forage in a single bite. Since the meal is not immediately chewed, ruminants can eat a large amount of forage in a short time period. The forage is then stored in the reticulorumen until the animal has some spare time. Later, while the animal is resting, the forage is regurgitated, chewed, and re-swallowed. This process called rumination (chewing a cud) serves to reduce feed particle size. Rumination also mixes the regurgitated forage with saliva. Saliva has a buffering effect on the rumen, which helps keep the pH near 7. This level of acidity is where the rumen "houseguest bacteria" are most comfortable. Starchy feeds, such as corn, can be digested by rumen bacteria, but some forage must be present in the diet to ensure proper rumen function.

Other than serving as a storage area for ingested forages, the reticulorumen serves several other

Animal Science Facts

Calves are not born with full functioning rumens. Their rumens gradually develop over the first few months of life.

purposes. The reticulorumen is a large fermentation vat containing billions of bacteria. These live-in guests digest complex carbohydrates that the animal's own digestive system cannot. The products of the fermentation process are known as **volatile fatty acids** or VFAs. VFAs are absorbed by the lining of the rumen and serve as an energy source for the animal. While the bacteria are digesting cellulose, they are multiplying, growing, and producing bacterial proteins from nitrogen contained within the forage. When chyme moves to the omasum and then into the abomasum, many bacteria are also moved along. When these bacteria are deposited in the abomasum, acid kills them. These dead bacteria provide the ruminant with large amounts of high-quality protein. Except in specially prepared diets, none of the protein entering the abomasum is the same as when it entered the animal. Bacteria disassemble and reassemble all of it.

Ruminant diets are typically forage-based. Forages include hay, silage, and pasture grasses or legumes. High-producing dairy cows and cattle being fed for slaughter also consume significant amounts of feed grains like corn.

The relationship between ruminants and bacteria is called a **symbiotic** relationship. Symbiotic relationships between two species are mutually beneficial. In this case, the cattle provide bacteria with food and an ideal environment in which to grow and reproduce. The bacteria provide ruminants with VFAs and microbial protein.

HORSES

Horses are also adapted to digesting high-forage diets, but in a slightly different way. See figure 4-13. The digestive tract of a horse is similar to that of a monogastric until the end of the small intestine. There, the cecum is enlarged and contains a population of bacteria, which ferment forages. The cecum lining absorbs VFAs, but microbial protein is lost in the feces. Horse diets are usually composed of hay and grains such as oats and corn. The proportion of grain in the diet depends on the horse's activity level. More active, harder-working horses need more energy in their diets.

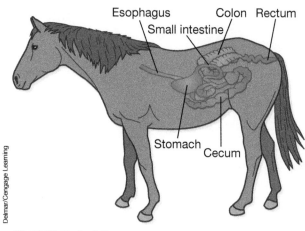

Delmar/Cengage Learning

FIGURE 4-13
The equine digestive system can digest forages, but employs a cecum rather than a rumen for the task.

SUMMARY

Animals must consume and digest food to produce animal products, such as meat, wool, milk, and eggs, and to do work. Six essential nutrients must be present in the diet: water, protein, carbohydrates, fats, vitamins, and minerals. Animals need all these to provide building materials and fuel for cellular growth and production. Not all energy consumed is available for production; some energy is lost in feces, urine, and to heat production. Feedstuffs are ingested by animals and are broken down into simpler compounds as they pass through the digestive system. These simpler compounds are absorbed into the bloodstream and are transported to the cells where they are reassembled into cellular products. Swine have a very basic digestive system that consists of the mouth, teeth, esophagus, stomach, small intestine, cecum, and large intestine. All other species have this same system with minor modifications. Poultry store undigested feed in the crop before it enters the true stomach and gizzard. Cattle and sheep have a special chamber in the stomach, called the rumen, which contains a population of microbes that digest complex carbohydrates. Horses have a population of microbes housed in the cecum, an appendage of the large intestine.

CHAPTER REVIEW

EXPERIENTIAL LEARNING OPPORTUNITIES

1. Investigate a career as an animal nutritionist. Locate a feed store in your area. If a nutritionist is employed there, ask to interview that individual or spend a day with him or her. Pay special attention to the computer programs that the nutritionist uses to balance rations.

2. Some country extension educators specialize in the animal sciences. Call your county extension office to see if there might be opportunities to work with a livestock agent. Also ask if summer employment opportunities may be available, either volunteer or paid positions.

DEFINE ANY TEN KEY TERMS

abomasum

amino acid

cecum

chyme

complex carbohydrates

crop

gizzard

hydrochloric acid

macrominerals

microminerals

monogastric

net energy

omasum

peptidases

peristalsis

proventriculus

reticulorumen

ruminant

silage

simple carbohydrates

symbiotic

villi

volatile fatty acids

QUESTIONS AND PROBLEMS FOR DISCUSSION

1.   What are the six essential nutrients in the diets of animals?

2.   True or False. Water accounts for 20 percent of total body weight.

3.   Protein consists of nitrogen containing building blocks called _____.

4.   Give two examples of simple carbohydrates.

5.   List the fat-soluble vitamins.

6.   List the water-soluble vitamins.

7.   Which vitamin helps to regulate blood clotting?

8.   True or False. Salt is a micromineral.

9.   Are chromium and cobalt macrominerals or microminerals?

10.   The major chemical element in all carbohydrates is _____.

11.   List two complex carbohydrates that simple-stomached animals cannot digest.

12.   Food moves through the esophagus by involuntary muscle contractions called _____.

13.   Which part of the poultry digestive tract grinds food?

14.   Name three different forages.

15.   What are the four parts of a ruminant stomach?

16.   Which class of organisms digests cellulose in a ruminant stomach?

17.   Which part of the four-part ruminant stomach traps foreign objects?

18.   Describe a symbiotic relationship.

19.   Compare the function of a horse's cecum to a cow's rumen.

20.   Contrast the use of microbial protein in cattle and horses.

5

Biology of Reproduction

GETTING IN THE FAMILY WAY!

Objectives

After completing this chapter, students should be able to:

▶ Identify and explain the functions of male and female reproductive tract

▶ Analyze physiological changes in the female during fertilization, gestation, parturition, and lactation

▶ Describe practical reproductive differences, such as signs of heat, length of estrus, and gestation period, among swine, cattle, sheep, and horses

Key Terms

capacitation	freemartin	testis
cervix	infundibulum	urethra
colostrum	Leydig cells	uterus
corpus hemorrhagicum	lordosis	vagina
corpus luteum	ovary	vas deferens
efferent ducts	oviduct	vitelline block
epididymis	penis	vulva
estrous cycle	rete testis	zona pellucida
follicle	seminiferous tubules	zygote

Career Focus

Antonio owned a small flock of Suffolk breeding sheep that he used for his SAE project in FFA. For two years in September, he borrowed a neighbor's ram to breed his ewes. After the 150-day gestation period, the ewes lambed in February, were weaned at the end of April, and spent from May to August loafing in the shade. Antonio thought his ewes could be more profitable if they were bred back as soon as they weaned their lambs. So the third year he borrowed his neighbor's ram in May and expected his ewes to lamb again in October. When October came, there were no lambs. Antonio could not understand why until his agriculture teacher taught a lesson on sheep reproduction when he learned that some sheep breeds can only become pregnant when daylight length is shortening. Antonio decided he wanted to become a reproductive physiologist with a goal of learning how to manipulate seasonal breeders to increase efficiency by reproducing at any time of the year.

INTRODUCTION

Efficient reproduction may be the single most important factor in the profitability of livestock production. Without baby pigs to sell, swine producers have no income. Beef cows must each raise a calf every year, or the business of beef production is unprofitable. Likewise, dairy

© Margo Harrison. Used under license from Shutterstock.com

FIGURE 5-1

Successful reproduction is imperative in most animal agriculture enterprises.

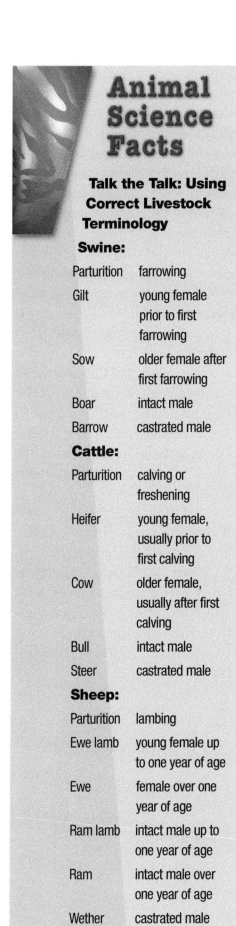

Animal Science Facts

Talk the Talk: Using Correct Livestock Terminology

Swine:

Parturition	farrowing
Gilt	young female prior to first farrowing
Sow	older female after first farrowing
Boar	intact male
Barrow	castrated male

Cattle:

Parturition	calving or freshening
Heifer	young female, usually prior to first calving
Cow	older female, usually after first calving
Bull	intact male
Steer	castrated male

Sheep:

Parturition	lambing
Ewe lamb	young female up to one year of age
Ewe	female over one year of age
Ram lamb	intact male up to one year of age
Ram	intact male over one year of age
Wether	castrated male

(continued)

cattle must have a calf to produce salable milk. The same theory holds true for sheep and, to some extent, horses. Although farm animals look very different, the biology of reproduction for all species is very similar.

ANATOMY OF REPRODUCTIVE TRACTS

To be able to understand events in the fertilization, gestation, and birthing process, you must first have a basic understanding of reproductive anatomy.

FEMALE REPRODUCTIVE SYSTEM

The primary reproductive organ of females is the **ovary** (see figure 5-2) where meiosis takes place and fertile eggs develop. Along with producing eggs, the ovary produces hormones at various times during the reproductive cycle.

Different structures are found on the ovary at different times during the reproductive cycle. A **follicle** is a developing egg that has not yet been released. Developing follicles produce a hormone called estrogen. Estrogen serves to further develop the follicle. When the follicle is mature, it ruptures (much like acne) and the egg is released. The release of the egg causes estrogen levels to fall. The blood clot remaining after the egg is released is called a **corpus hemorrhagicum** (CH) or bloody body. The CH lasts for a few days after which a yellow body begins to form at the site of the egg release. This new structure is called a **corpus**

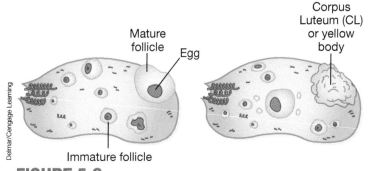

FIGURE 5-2

The ovary on the left contains a mature follicle while the ovary on the right contains a corpus luteum (CL).

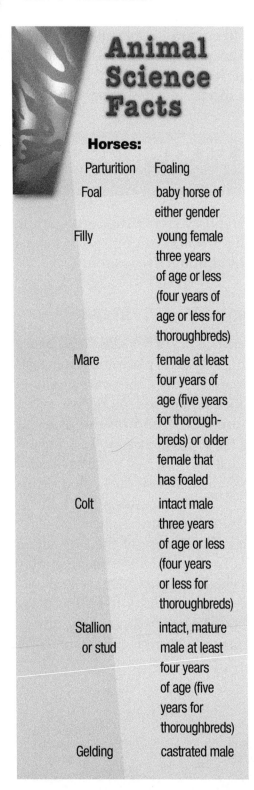

Animal Science Facts

Horses:

Parturition	Foaling
Foal	baby horse of either gender
Filly	young female three years of age or less (four years of age or less for thoroughbreds)
Mare	female at least four years of age (five years for thorough-breds) or older female that has foaled
Colt	intact male three years of age or less (four years or less for thoroughbreds)
Stallion or stud	intact, mature male at least four years of age (five years for thoroughbreds)
Gelding	castrated male

luteum (CL). The CL produces a hormone called progesterone, which sustains pregnancy. Progesterone also prevents a new follicle from developing. If pregnancy occurs, the CL will persist until just before the fetus is delivered. At that time, the CL will regress. If pregnancy does not occur, another hormone secreted by the **uterus** (womb) called prostaglandin causes the CL to regress and allows a new follicle to begin developing. Then, the cycle repeats.

This cycle of structures on the ovary corresponds with what is called the **estrous cycle**. The period during which a follicle is developing is called proestrus. The time when the egg is released is called estrus. It is during estrus that the female is receptive to be mated by a male. This period is also known as standing heat. The next period is called metestrus. During this period, the CL is actively producing progesterone if the female is pregnant. If the female is not pregnant, she goes into the next phase called diestrus during which the CL recedes and a new follicle develops. If the female is pregnant, the CL remains active until the end of the pregnancy. Before puberty and for a period of time after giving birth, females do not cycle at all. Noncycling females are said to be anestrus.

During, or around estrus or standing heat, the egg is released into a funnel called the **infundibulum**, through which the egg moves into a tube called the **oviduct**. See figure 5-3. Fertilization of eggs with sperm takes place within the upper portions of the oviduct. After fertilization, the egg continues moving toward the uterus where it attaches to the uterine lining and develops until birth.

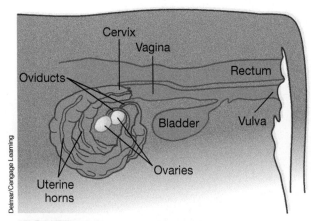

Delmar/Cengage Learning

FIGURE 5-3

Structures in the reproductive tract of a cow.

The upper portion of the uterus consists of two horns, one each leading to the oviduct and the ovary, respectively. The lower portions of the uterine horns attach to form the body of the uterus. The size and length of the uterine horns vary among species. Embryos can attach in both horns in swine, but they normally attach only in the horn closest to the ovary of origin in cattle, sheep, and horses. In species where single births are common, eggs are released from alternate ovaries in each cycle.

First in the path leading from the uterus to the outside of the animal is the **cervix**. In some species, sperm is deposited in the cervix. The cervix also covers the uterus after pregnancy has been established to guard against bacteria infecting the developing fetus. The next structure toward the outside of the female is the **vagina**, which serves as a site of insemination in some species. The visible part of the female reproductive tract is called the **vulva**.

MALE

The primary male reproductive organ is the **testis** (plural is testes). See figure 5-4. Meiosis takes place in **seminiferous tubules** buried deep within the testicle tissue. See figure 5-5. Surrounding the seminiferous tubules are **Leydig cells**, which produce the hormone testosterone. Testosterone is critical to the normal development of sperm cells. Immature sperm collect in ducts called **rete testis**, and then travel to a bigger

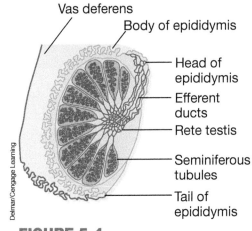

FIGURE 5-4

The testis is the primary reproductive organ of the male.

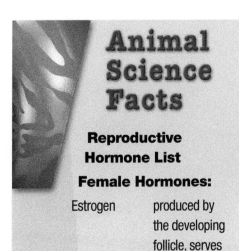

Animal Science Facts

Reproductive Hormone List

Female Hormones:

Estrogen	produced by the developing follicle, serves to further develop the follicle
Progesterone	produced by the corpus luteum, sustains pregnancy, and inhibits new follicle development
Prostaglandin	produced by the uterus in the absence of pregnancy, causes corpus luteum to regress, and allows a new follicle to develop on the ovary.
Relaxin	produced by the corpus luteum, causes ligaments around the birth canal to relax during parturition

(continued)

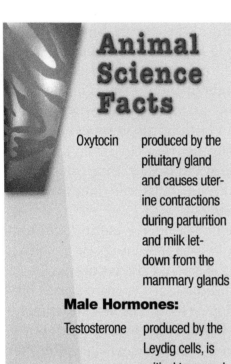

Animal Science Facts

Oxytocin   produced by the pituitary gland and causes uterine contractions during parturition and milk let-down from the mammary glands

Male Hormones:

Testosterone   produced by the Leydig cells, is critical to normal sperm development, and causes secondary sexual characteristics such as a cresty neck in bulls.

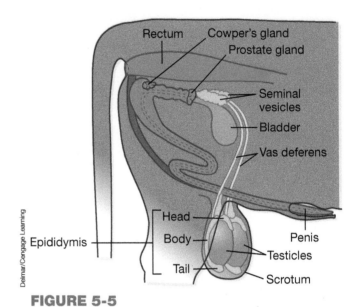

FIGURE 5-5

Structures in the reproductive tract of a bull.

storage area in the **efferent ducts**. The final sperm collection site is the **epididymis**, which has a head, a body, and a tail. Sperm continue the maturation process as they travel through the epididymis. During ejaculation, the sperm travel from the epididymis, through the **vas deferens**, then to the **urethra** (these connect bladder and vas deferens to penis opening), and **penis** (male copulatory organ) before being deposited in the female. During this journey, several sets of glands (Cowper's gland, prostate, and seminal vesicles) add fluids to the sperm. These fluids supply nutrients for the sperm to live after entering the female reproductive tract. Sperm cells mixed with these fluids are called semen.

THE REPRODUCTIVE CYCLE

Now that you know about the reproductive anatomy of males and females, we will discuss fertilization, pregnancy, the birthing process, and lactation.

FERTILIZATION

During mating, sperm are deposited and swim up the female reproductive tract with the aid of female uterine contractions, while undergoing a process called **capacitation**. During capacitation, fluids present in the

Animal Science Facts

Superovulation and embryo transfer allows a producer to get several offspring from one cow each year. A series of hormone shots stimulate the donor cow to ovulate more than the normal single egg each cycle. The donor is then artificially inseminated. About seven days later, the uterus is flushed with fluid. The flushing process washes out all the fertilized eggs, which are collected, evaluated for quality, and either immediately implanted into recipient cows or frozen for future use. The process was first accomplished in the commercial cattle industry in the 1970s but now is a common way for producers to increase the genetic impact of an outstanding female.

female reproductive tract wash the sperm and change their chemical composition, making them capable of fertilization. When sperm reach the upper portions of the oviduct, where fertilization takes place, they meet with the recently shed egg. The egg is surrounded by a barrier called the **zona pellucida**. Multiple sperm cells attach to the zona pellucida and attempt to gain entrance to the egg. However, only a single sperm can penetrate the egg. Once a single sperm has penetrated the zona pellucida, the chemical composition of the zona pellucida changes, and no more sperm can enter. The technical name for this process is the **vitelline block**. See figure 5-6. After the egg has accepted a sperm, the nuclei of both gametes mix, and the diploid fertilized egg, called a **zygote**, is formed.

The zygote travels the remaining distance to the uterus where it is nourished by uterine fluids. The single-cell zygote divides to 2, 4, 8, then 16 cells. See figure 5-7. During these early stages of development, the zygote becomes an embryo. The embryo soon attaches to the side of the uterus and develops an attachment to the female's blood supply through the placenta and umbilical cord. During early cell division, until the internal organs are formed, the unborn young is known as an embryo. Later in the developmental process, the embryo is known as a fetus.

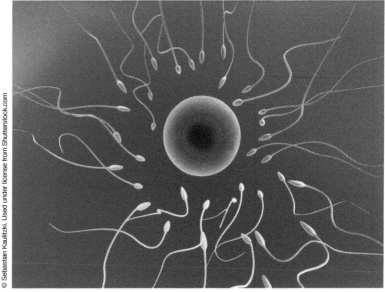

© Sebastian Kaulitzki. Used under license from Shutterstock.com

FIGURE 5-6

Although many sperm arrive at the egg, only one sperm can enter due to a process called the Vitelline Block.

8-Cell embryo

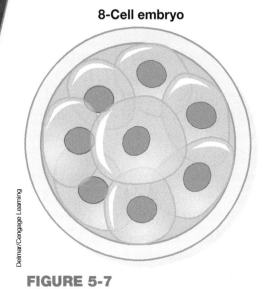

Delmar/Cengage Learning

FIGURE 5-7

An 8-cell embryo.

PREGNANCY

During the period the embryo or fetus is developing in the uterus, no new follicles develop in the ovary. A thick mucous plug forms within the cervix to prevent bacteria from contacting the developing fetus.

PARTURITION (BIRTH)

When the pregnancy nears full term, the fetus signals to the mother when parturition will take place. The fetus starts to secrete a hormone, which causes the mother to release prostaglandin. In turn, prostaglandin ruptures the CL, ending the pregnancy, and causing the fetus to be born. The hormone released by the fetus also causes the mother to release two other hormones. The first is called relaxin, which relaxes the ligaments around the birth canal and allows the young to pass through during delivery. See figure 5-8. The second is called oxytocin,

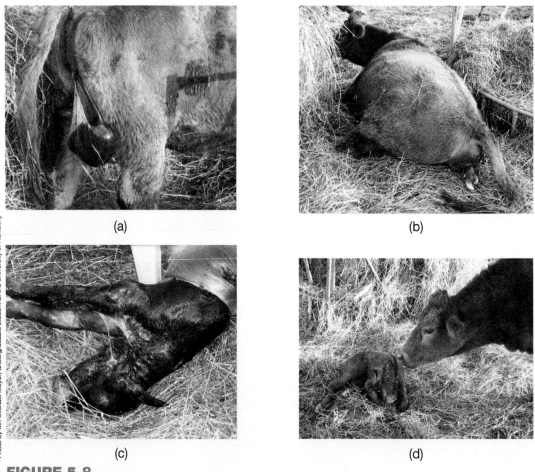

Photo by Mr. Michael Meyer, undergraduate student at the University of Kentucky

FIGURE 5-8

Relaxin allows ligaments surrounding the birth canal to stretch enabling the dam to eject the fetus.

which causes milk let-down from the mammary gland and uterine contractions to expel the fetus.

LACTATION

Milk let-down is controlled by the hormone oxytocin. Milk formation in the mammary gland is controlled by the hormone prolactin. The first milk available from the female to newborn young is called **colostrum**. Colostrum is thicker and richer than the milk that is produced later in lactation. Colostrum is produced in small amounts. It contains antibodies to any diseases the mother has contracted and protects the newborn. Newborn animals need colostrum within 24 hours after birth (preferably within six hours) because after that time antibodies are much less readily absorbed.

Later lactation is characterized by frequent milk let-down. The amount of milk produced by the female changes as the newborn grows. As the young animal grows, more and more milk is produced to compensate for its increasing appetite. At some point, the female is producing as much milk as her genetics, nutrition, and mammary gland will allow. This point is called the "peak of lactation." After the lactation peak, milk production gradually drops off and eventually stops.

In the real world of livestock production, producers usually stop lactation by weaning or removing the young animal from the mother. If young animals or milking machines do not remove milk from the mammary gland, milk production will naturally cease.

The mammary gland is amazing in its ability to capture nutrients from the bloodstream and synthesize large quantities of milk. See figure 5-9. High-producing dairy cattle routinely produce over 100 pounds or 200 half-pints of milk per day. To produce this much milk, cows have to eat and drink a lot. Lactating dairy cattle eat about 3.5 percent of their body weight each day, not including water. That's about 50 pounds of dry feed. Many females cannot physically eat enough to maintain body weight during peak lactation and must utilize body reserves of fat, protein, calcium, and other minerals to maintain milk production. This body weight loss can delay cyclic activity and rebreeding in some high-milking females.

Animal Science Facts

Blood tests can easily and quickly confirm pregnancy in cattle as early as 30 days after breeding. The Biopryn® test measures for the presence of a protein produced by the developing fetus that circulates in the dam's bloodstream. Producers can quickly and inexpensively collect a blood sample from a mated animal, mail it away, and receive a pregnancy diagnosis within a few days. The process is much less expensive than hiring a veterinarian to perform rectal palpation to confirm pregnancy and nonpregnant animals can quickly be returned to an aggressive estrus synchronization protocol.

FIGURE 5-9

An elite dairy cow may secrete over 200 pounds (about 25 gallons) of milk per day during peak lactation.

REPRODUCTION IN SWINE

Swine display standing heat at approximately 21-day intervals. Heat can be determined by observing a swollen vulva or by placing weight on the female's back while rubbing the flank. If the sow or gilt does not try to run away, she is in standing heat. See figure 5-10. In addition, sows may vocalize, pace, and try to mount other pigs. The ears may pulse up and down when pressure is placed on the back, further confirming standing heat. This response is called **lordosis**. The presence of a boar during heat checking results in a stronger expression of standing heat. Gilts usually display first estrus at five to six months of age. Female swine will also display estrus throughout the year because they are not seasonal breeders.

Swine should be mated at least twice, with the first service timed 12 hours after the beginning of standing heat. See figure 5-11. Subsequent matings should be spaced at 12- or 24-hour intervals until the sow or gilt will no longer stand for mating. Estrus normally lasts 24 to 72 hours. Swine can be bred either with a boar or artificially using frozen or freshly collected semen.

Pregnancy in swine can be checked ultrasonically 30 days after breeding, or by watching for the

FIGURE 5-10

A sow in heat will stand rigidly when weight is placed on her back.

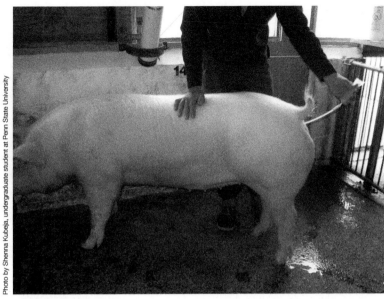

Photo by Shenna Kubeja, undergraduate student at Penn State University

FIGURE 5-11

Most commercial sows are mated artificially using fresh or frozen semen.

absence of signs of heat 21 days after breeding. Gestation length for swine is normally between 112 and 118 days, averaging about 114 days. Swine normally ovulate 8 to 20 ova per estrus. Most ova are fertilized, but several normally die before attaching to the uterus. The number of eggs that attach to the uterus is somewhat dependent on the length of the uterus; the longer the uterus, the more the eggs that can attach. Therefore, older, larger sows normally have larger litters because more embryos can attach the uterus.

Swine normally give birth to, or farrow, litters averaging 9 or 12 pigs, but litter size is highly variable. See figure 5-12. Sows nurse their young until removal at weaning, normally at three to six weeks of age in most management situations. If not weaned earlier, sows will stop lactating after eight to nine weeks. Sows will usually return to estrus four to seven days after weaning.

REPRODUCTION IN CATTLE

The normal estrous cycle in cattle is 21 days. In other words, barring pregnancy, an egg is released into the oviduct every three weeks. Heifers reach puberty (first estrus) at about 9 to 10 months of age and should

FIGURE 5-12

Nationally, commercial sows wean about ten pigs per litter.

FIGURE 5-13

Cows in heat will try to mount other cows.

be bred at approximately 15 months to have their first calf at two years of age. Cattle will display estrus during all seasons of the year.

Shortly before the release of the egg, cattle display standing heat, which means they will stand still if another cow (or bull) tries to mount them. Standing heat is the only positive sign that a cow is in heat. However, many cows will vocalize, try to ride other cows, and act restless when nearing standing heat. See figure 5-13. Producers should routinely check for signs of heat twice daily. Early morning and evening

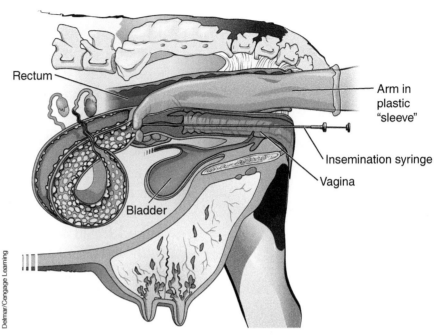

Delmar/Cengage Learning

FIGURE 5-14

Cattle have been mated artificially with frozen semen since the 1930s.

are good times to check for heat since cattle may not be active at midday.

Many cattle are bred using artificial insemination with frozen semen. See figure 5-14. This allows producers to select sires of superior genetic quality without the cost of maintaining bulls. The ideal time for insemination in cattle is during the last half of standing heat, which lasts 8 to 18 hours. Cattle are usually inseminated only once toward the end of, or shortly after the standing-heat period. Cows display heat about 40 to 60 days after calving and normally become pregnant while still lactating.

Pregnancy in cattle is ideally confirmed by a rectal examination, called palpation, about 40 to 50 days after breeding, or by an inexpensive blood test that checks for a specific protein produced by the fetus after 30 days. Otherwise, the absence of heat signs 21 days after breeding can indicate a pregnancy. The average gestation length for cattle is 280 to 285 days or about nine and a half months. Approaching birth in cattle, called calving or freshening, can be signaled by restlessness, frequent urination, relaxation of ligaments around the tail head, presence of milk, and loosening of the vulva.

Cattle normally give birth to a single offspring, although twins are fairly common. Triplets are rare. In a split set of twins (one bull and one heifer), the heifer will be sterile 80 to 90 percent of the time. This phenomenon occurs almost solely in cattle but has been observed in sheep also. This condition is due to the passing of hormones from the male to the female fetus, which inhibits the growth and development of the female reproductive tract. This sterile heifer born twin to a bull is known as a **freemartin**.

Dairy cattle are normally dried off or left unmilked for about 60 days before the birth of a calf to allow them to gain body weight. Beef calves are normally weaned at four to eight months of age. After weaning, beef cows remain dry until the next calf is born.

REPRODUCTION IN SHEEP

Ewes have a somewhat shorter estrous cycle than cattle and swine, averaging 16 days instead of 21. Detection of estrus is very difficult in sheep. Standing for the ram to breed is the best indication of estrus. This difficulty limits the use of artificial insemination in sheep. Some producers use a vasectomized ram with a chalking device attached to his chest to mark ewes in heat. See figure 5-15. These ewes can immediately be bred artificially.

Some breeds of sheep are known as seasonal breeders. In general, black-faced breeds, such as Hampshires or Suffolks, are the most seasonal, showing estrus only in the fall months as day length shortens. White-faced sheep are less seasonal and may show estrus at any time during the year. Ewe lambs experience their first estrus at six to nine months of age. Furthermore, ewe lambs may be bred in their first fall so that they will lamb when they are yearlings; however, they will give birth to a greater percentage of single lambs than older ewes do. Older ewes have a high incidence of twins and triplets, with white-faced sheep generally averaging more lambs per year than black-faced sheep.

Photo by Dr. Jay Daniel, Berry College

FIGURE 5-15

A marking harness placed on a vasectomized ram marks ewes in heat.

FIGURE 5-16

Most breeds of sheep typically give birth to one to three lambs.

Pregnancy can be determined ultrasonically or by palpation at 40 to 60 days after mating. Producers normally check pregnancy by observing for estrus 16 days after mating. If the ram does not breed the ewe again, she is determined to be pregnant. The average gestation length for sheep is about 150 days, or five months.

Ewes normally give birth to one to three lambs. See figure 5-16. Four lambs are fairly common with very prolific breeds, such as Finnsheep.

Lambs may be weaned anytime after they start eating feed on their own. Most management strategies call for only one lamb crop per year. However, using white-faced sheep that will breed out of season, one can possibly produce a lamb crop every eight months.

REPRODUCTION IN HORSES

Estrus detection in horses requires an intact male (stallion), and is known as teasing. See figure 5-17. When a teaser stallion is given visual contact to a mare in heat, the mare will assume a squatting position, urinate frequently, and display winking of the vulva. Heat detection can also be aided by palpation of the ovaries for mature follicles.

Horses are naturally seasonal breeders, showing estrus every 21 days during the period from mid-April to mid-September. Controlling the lighting in a box

FIGURE 5-17

Stallions exhibit the Flehman response when determining if a mare is near estrus.

stall can fool the mare into displaying estrus at other times of the year.

Young female horses (fillies) normally show their first estrus at 12 to 15 months of age, but breeding is normally delayed until the filly is two or three years old. Estrus may last from three to seven days, and ovulation occurs one to two days before the end of estrus.

Horses can be mated either naturally with a stallion or artificially. However, some breed associations either will not allow registry or will severely restrict the registration of horses sired by artificial insemination.

Pregnancy can be determined by the absence of the next heat period, rectal palpation, ultrasound examination, or blood test. The normal gestation period for a horse is about 330 days or 11 months.

Mares normally show heat seven to ten days after foaling. This heat is known as foal heat. Mares may be rebred during foal heat or during the next heat period 20 to 25 days later. Mares normally rebreed while lactating. Foals are normally weaned in the fall of the year.

SUMMARY

The primary reproductive organ of the female is the ovary. Ovaries produce eggs as well as hormones that allow and sustain pregnancy. Eggs are fertilized in the oviduct, then travel to the uterus where attachment occurs. Fertilization can only occur shortly after ovulation at a certain point in the estrous cycle called estrus or standing heat. The primary male reproductive organ is the testis. Sperm are manufactured in the testis and are transferred through a system of ducts to the epididymis. During mating, sperm are mixed with other fluids and ejaculated into the female reproductive tract through the penis. Sperm and egg meet at fertilization. Only one sperm can fertilize an egg. All others are blocked from entrance. The fetus signals the female when parturition or birth will occur. The first milk is called colostrum. Later in lactation, milk production peaks then drops until weaning. There are minor differences in the reproductive systems and reproductive performance of swine, cattle, sheep, and horses.

CHAPTER REVIEW

EXPERIENTIAL LEARNING OPPORTUNITIES

1. A veterinarian can teach students how to palpate cattle for pregnancy. Contact a veterinarian to see if this could be arranged.

2. Consider starting a breeding project for an SAE. Spend time considering facility requirements and economics before beginning.

DEFINE ANY TEN KEY TERMS

capacitation	oviduct
cervix	penis
colostrum	rete testis
corpus hemorrhagicum	seminiferous tubules
corpus luteum	testis
efferent ducts	urethra
epididymis	uterus
estrous cycle	vagina
follicle	vas deferens
freemartin	vitelline block
infundibulum,	vulva
Leydig cells	zona pellucida
lordosis	zygote
ovary	

QUESTIONS AND PROBLEMS FOR DISCUSSION

1. Which hormone is produced by follicles developing on the ovaries?

2. What does CL stand for with reference to the estrous cycle?

3. Describe the four stages of estrus.

4. List the organs of the female reproductive tract.

5. Give another name for the womb.

6. Where, in the male reproductive tract, is testosterone produced?

7. Explain why no more than one sperm can penetrate a single egg.

8. Where does fertilization occur?

9. Into which female structure does the fertilized embryo implant?

10. Which hormone triggers milk let-down?

11. Explain the signs of standing heat in cattle.

12. Which two species discussed in this chapter can be seasonal breeders?

13. How long is the estrous cycle for swine, cattle, sheep, and horses?

14. How long is the gestation period for swine, cattle, sheep, and horses?

15. At what age do swine, cattle, sheep, and horses experience their first estrous period?

16. Why are dairy cattle dried off two months prior to calving?

17. How many half-pint sized milk cartons come from 100 pounds of milk?

18. Do older or younger ewes tend to have more incidents of multiple births?

19. A stallion used to detect mares in heat is called a _____.

20. At what time of year are foals usually weaned?

CHAPTER

6

Genetics

A CHIP OFF THE OLD CHROMOSOME

Objectives

After completing this chapter, students should be able to:

▶ Describe how the gender of offspring is determined

▶ Explain how genotype and phenotype are different

▶ Differentiate between qualitative and quantitative inheritance

▶ Identify breeding systems commonly used in commercial animal agriculture

▶ Explain why heritability, heterosis, selection intensity, and generation interval are important to genetic change

▶ Discuss how genomics can increase the rate of genetic change

Key Terms

alleles	genomics	incomplete dominance
codominance	genotype	line breeding
complete dominance	heritability	out breeding
contemporary groups	heterosis	phenotype
crossbreeding	heterozygous	qualitative traits
generation interval	homozygous	quantitative traits
genetics	inbreeding	selection intensity

Career Focus

Allison has been reading about genetic selection in her Animal Science textbook. She understands most of what she reads about selection intensity, generation interval, and genomics, but still has some questions. At a trip to a local grocery store, Allison spots a magazine next to the checkout counter touting the headline, "New Genomic Technology Threatens Safety of World Food Supply." While Allison is not completely familiar with the details of genomics, she cannot quite understand how this technology would endanger people. She remembers several other recent magazine articles that have expressed fear of new and innovative technologies. Allison decides she needs to be more familiar with the technologies that make agriculture more efficient. Since she's always had a gift for writing, maybe she could land a job as a technology writer and someday explain, in simplified, layman's language, how new, science-based expertise helps to safely feed the world.

INTRODUCTION

Animal selection based on **genetics** (the study of heredity) can be traced to two insightful individuals. The first was Gregor Mendel, an Austrian monk who is known as the father of genetics. He discovered that factors, known as genes, pass from one generation to the next that influence physical characteristics. The second was Robert Bakewell, an Englishman who is known as the father of modern animal breeding.

© Candor 36. Used under license from Shutterstock.com

FIGURE 6-1

Genetic inheritance explains why offspring resemble their parents.

Bakewell demonstrated that if a male and female that excelled in a certain trait were mated, the resulting offspring also excelled in the same trait. The science of genetics has broadened greatly since the primitive efforts of these two forefathers; however, the basic premises remain the same.

MENDELIAN GENETICS

Gregor Mendel was the first to discover that traits were inherited through units called genes that are present on chromosomes. He also deduced that genes are present in pairs with half of each pair being inherited from each parent. Each gene could then be independently transmitted, unchanged, to the next generation. Pairs of genes are known as **alleles**. Figure 6-2 illustrates a pair of alleles on homologous chromosomes. Remember, chromosomes are made of DNA and are found in the nucleus of cells.

FIGURE 6-2

Chromosomes are X-shaped structures occurring in pairs. Genes coding for the same trait occupy the same space on each chromosome.

DETERMINATION OF GENDER

As discussed in Chapter 1, when fertilization takes place, genetic materials from the gametes (sperm and egg), fuse together to form a fertilized zygote that inherits genetic material (genes) from both parents and thus will display characteristics from both. The determination of the sex of the zygote depends on a pair of chromosomes called the sex chromosomes. Male sex chromosomes can be either X or Y. All eggs from the female receive an X chromosome. A zygote that gets an X chromosome from the sperm will be female. A zygote that receives a Y chromosome from the sperm will be male. So, a

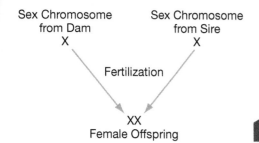

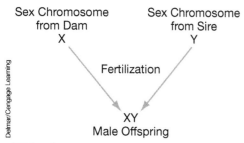

FIGURE 6-3

Offspring gender depends solely on which sex chromosome is inherited from the sire.

female zygote will have two X chromosomes (XX) and a male zygote will have an X and a Y chromosome (XY). See figure 6-3.

GENOTYPE AND PHENOTYPE

The **genotype** of an animal is its actual genetic code. The animal's genotype controls physical traits, such as color, as well as potential performance traits, such as average daily weight gain. Genotype is the actual genetic code that cannot be changed by environmental factors. **Phenotype** is the part of the genotype the animal expresses. In some cases, like in coat color, phenotype cannot be changed by the environment. In other cases, it can. For example, a lamb may have the genotype to gain one pound per day; however, because it receives an inferior diet, the lamb only gains one-half pound per day. The lamb's phenotype for daily gain is one-half pound per day. In this case, an environmental factor (the inferior diet) hides the true genetic potential (genotype) of the lamb.

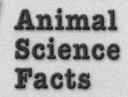

Animal Science Facts

The X chromosome is significantly larger and heavier than the Y chromosome. This allows semen to be separated into fractions that contain mostly X - carrying sperm cells. Sexed semen can be frozen and used to produce a very high percentage of female offspring. This technology is very useful to dairy producers who have little use for male calves.

QUALITATIVE TRAITS

Qualitative traits are traits controlled by only a single pair of genes and cannot be altered by the environment. They are sometimes called either/or traits because their phenotype is either one thing or the other. Coat color is a good example of a qualitative trait. If a cow is genetically programmed to be black, no environmental factor can change the fact that the cow is black. Other examples of qualitative traits include the polled or horned conditions of cattle and sheep and blood type of all species.

Qualitative traits most easily show how genes are inherited. Coat color in Shorthorn cattle is a classic example. Traditionally Shorthorns had three phenotypes: white, red, and roan (a mixture of white and red hairs). See figure 6-4. The two possible genes are red (designated R) and white (designated r). A red Shorthorn has two copies of the red gene (genotype

Photo courtesy of Dave McElhaney

FIGURE 6-4

Roan Shorhorns have a mixture of red and white hairs.

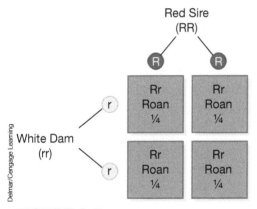

Delmar/Cengage Learning

Red Sire
(RR)

White Dam
(rr)

FIGURE 6-5

Punnett square of RR (red) × rr (white) mating in Shorthorn cattle. Each box represents a possible genotype of the offspring. All offspring are Rr (roan).

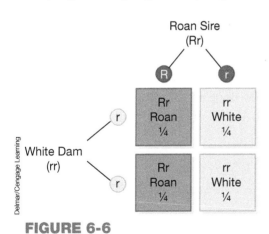

Delmar/Cengage Learning

Roan Sire
(Rr)

White Dam
(rr)

FIGURE 6-6

Punnett square of Rr (roan) × rr (white) mating in Shorthorn cattle. On average, half the offspring will be Rr (roan) and half will be rr (white).

designated by RR). A white Shorthorn has two copies of the white gene (genotype designated by rr). A roan animal has one copy of each gene (genotype designated by Rr).

Coat color in Shorthorn cattle is an example of a qualitative trait with **incomplete dominance**. This means that both the R and the r gene are expressed. Neither gene completely masks the other's presence. Both are expressed to some degree (both red and white hairs are present with the Rr genotype). If a red Shorthorn bull (RR) is mated to a white Shorthorn cow (rr), all resulting offspring will receive one copy of each gene from each parent and will be roan (Rr). See figure 6-5. However, when **codominance** occurs, a blending of the allele pair is expressed. If codominance occurred in the Shorthorn example, the offspring would be pink! Punnett squares aid in determining the genotype of resulting offspring. If a roan bull is mated to a white cow, half of the resulting offspring will be white and half will be roan. See figure 6-6.

A qualitative trait with **complete dominance** can be illustrated by the coat color in Angus cattle. When dominance is complete, one gene completely masks the presence of the other. The two possible genes for coat color in Angus cattle are black (B) and red (b). If an animal has either one (Bb) or both (BB) copies

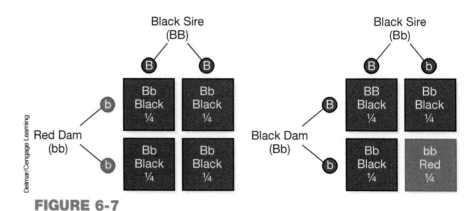

FIGURE 6-7

Punnett square for inheritance of coat color in Angus cattle. The black gene (B) is dominant and masks the red gene (b).

of the black gene, phenotypically it will be black. Only if an animal has both copies of the red gene (bb) will the animal be red. Figure 6-7 displays the inheritance of coat color in Angus cattle.

Animals that have two copies of the same gene are called **homozygous** for that trait. An Angus cow with the genotype BB for coat color would be referred to as homozygous black. Since black is dominant to red in Angus cattle, genotype BB could be called homozygous dominant. Genotype bb would then be called homozygous recessive. Animals that have one copy of each gene are called **heterozygous** for that trait. An Angus cow with the genotype Bb would be called heterozygous black. Animals that are homozygous have a 100 percent chance of passing that gene to their offspring. Therefore, if an Angus bull is homozygous black (BB), all of its offspring will be black. However, if an Angus bull is heterozygous black (Bb) and is mated to a heterozygous cow, one-fourth of the offspring (on average) will be red. In other words, in a mating of two heterozygous black Angus cattle, the chance of getting a red calf is one in four.

QUANTITATIVE TRAITS

Quantitative traits are controlled by several or many pairs of genes. Most economically important livestock traits such as growth rate, backfat depth, speed, milk production, and wool production are quantitative

Animal Science Facts

Lethal genes are rare, recessive genes that cause death of the fetus if present in the homozygous form. Parents can carry one copy of the lethal gene in the heterozygous form and be completely normal in appearance and performance. However, if two carrier parents each pass the recessive lethal gene to the same offspring, death occurs prior to birth or shortly thereafter. Examples of lethal genes are bulldog syndrome in cattle (calf is born with shortened legs and head), hydrocephalus in humans (water on the brain), and atresia coli (closed colon in horses). Often, new lethal genes are revealed when a carrier sire becomes very popular and breeders attempt to concentrate his genes through inbreeding. Three newly discovered lethal beef cattle traits fit this description: tibial hemimelia (TH) in Shorthorns, pulmonary hypoplasia with anasarca (PHA) in Maine Anjou cattle, and Arthrogryposis multiplex (AM) in Angus Cattle.

traits. The expression of quantitative traits can be influenced by environmental factors. Instead of the either/or inheritance of qualitative traits, quantitative traits are expressed across a range. For instance, in pigs, average daily weight gain is a quantitative trait that may range from zero to three pounds per day. The expression of average daily gain is a combination of genes and environment. Even though these traits are controlled by many pairs of genes, some animals have more desirable genes for a trait than other animals. Animals that are homozygous for these desirable genes have a better chance of passing those traits to their offspring. See figures 6-8 and 6-9.

To unmask the true genetic potential of an animal, it must be given an ideal environment. This means any obstacle that could cause the animal to express less than its genotype would allow under ideal conditions should be eliminated. Some of these environmental factors include nutrition, disease, temperature, and competition from other animals. Unfortunately, total elimination of these factors is nearly impossible. Therefore, livestock producers identify animals with superior genes for quantitative traits by comparing expression of those traits among animals that are fed and managed in the same way. These similarly fed and managed groups of animals are called **contemporary groups**. See figure 6-10.

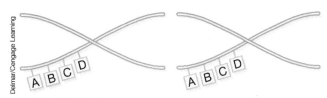

FIGURE 6-8

Chromosome representation of an animal that is homozygous for a quantitative (multiple gene) trait.

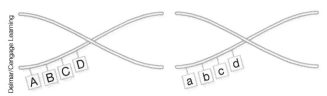

FIGURE 6-9

Chromosome representation of an animal that is heterozygous for a quantitative (multiple gene) trait.

Photo Courtesy of Robert Meinen

FIGURE 6-10

Animals of the same age fed and managed alike are called contemporary groups.

BREEDING SYSTEMS

Although genetic manipulation of animals in a laboratory is a recent scientific breakthrough, livestock producers have manipulated genes through inbreeding and crossbreeding for many years. Regardless of the method, most genetic manipulation is an effort to improve quantitative traits of livestock.

INBREEDING

Inbreeding is the mating of closely related animals, such as parent–offspring or brother–sister. See figure 6-11 for an example of an inbred pedigree. The concentration of genes resulting from inbreeding serves to make animals more homozygous for all traits, both qualitative and quantitative. For instance, mating a sire that has genes for outstanding performance with his daughters increases the probability that the resulting offspring will possess a greater proportion of the original sire's genes. One drawback with inbreeding is that both undesirable and desirable genes are concentrated in inbred animals. Any genetic flaw in the original sire would also be magnified in the offspring of a father–daughter mating. **Line breeding** is the mating of related animals that are not immediate family members.

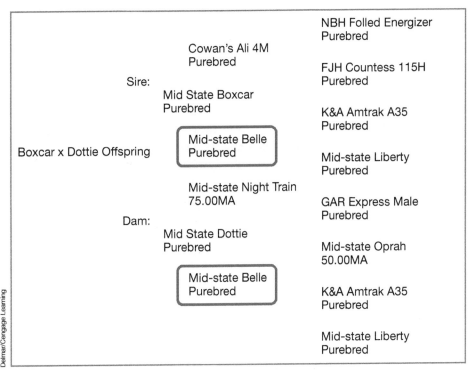

Boxcar x Dottie Offspring

Sire:

Mid State Boxcar
Purebred

Cowan's Ali 4M
Purebred

NBH Folled Energizer
Purebred

FJH Countess 115H
Purebred

Mid-state Belle
Purebred

K&A Amtrak A35
Purebred

Mid-state Liberty
Purebred

Mid-state Night Train
75.00MA

GAR Express Male
Purebred

Dam:

Mid State Dottie
Purebred

Mid-state Oprah
50.00MA

Mid-state Belle
Purebred

K&A Amtrak A35
Purebred

Mid-state Liberty
Purebred

FIGURE 6-11

Mid-State Belle appears as both the maternal and paternal grand dam in this inbred pedigree.

CROSSBREEDING (OUT BREEDING)

Crossbreeding or **out breeding** is the mating of animals that are not related. Crossbred animals have a higher proportion of heterozygous gene pairs than do non-crossbred animals. Offspring resulting from crossbred matings receive very different genes from the sire and dam because the parents are not related. The main advantage of out breeding is the increase in a phenomenon called **heterosis** in crossbred animals. See figure 6-12. Heterosis (sometimes called hybrid vigor) is the increase in a performance trait that exceeds the average of the parents. For instance, cattle breed A may have an average weaning weight of 450 pounds and cattle breed B may have an average weaning weight of 500 pounds. It would make sense that since half the genes for weaning weight come from the sire and half from the dam, the average weaning weight of the crossbred progeny would be 475 pounds. But the crossbred progeny of breeds A and B may display an average weaning weight of 525 pounds. This difference between the actual weaning weight (525 pounds) and the expected weaning weight (475 pounds) is a result of heterosis.

Trait	Percent Heterosis Advantage over Purebred		
	First Cross Purebred Sow	Multiple Cross Crossbred Sow	Crossbred Boar
Reproduction			
Conception rate	0.0	8.0	10.0
Pigs born alive	0.5	8.0	0.0
Litter size 21 days	9.0	23.0	0.0
Litter size weaned	10.0	24.0	0.0
Production			
21-day litter weight	10.0	27.0	0.0
Days to 220 lb	7.5	7.0	0.0
Feed/gain	2.0	1.0	0.0
Carcass composition			
Length	0.3	0.5	0.0
Back fat thickness	−2.0	−2.0	0.0
Lion muscle area	1.0	2.0	0.0
Marbling score	0.3	1.0	0.0

Delmar/Cengage Learning

FIGURE 6-12

Heterosis estimates for swine. Notice that highly heritable traits exhibit low heterosis.

HERITABILITY

As previously discussed, the true genetic potential of an animal can be altered or masked by many environmental factors. Geneticists have developed a system for separating genetics from environment, following this simple equation:

genetics + environment = genetic expression

The genetic expression is a quantitative trait, such as butterfat percentage, backfat, or wool production that we can measure. The genetics and environment parts of the equation must be estimated by genetic breeding experiments. Geneticists have separated the proportion of many different traits that result from genetics from the proportion that results from the environment. The proportion that results from genetic differences is called **heritability** and is expressed as a percentage.

For example, the heritability of reproductive traits, such as pigs born per litter, is quite low—only about 15 percent. This means that only 15 percent of the difference seen in pigs born per litter is a result of genetics. The other 85 percent results from environmental

Animal Science Facts

Although estimates vary, the following traits are listed with approximate corresponding heritability (percentage of observed trait performance resulting from genetics).

Number of live pigs farrowed	15%
Backfat thickness in swine	45%
Birth weight in beef cattle	40%
Carcass tenderness in beef cattle	60%
Mastitis susceptibility in dairy cattle	25%
Milking ability in dairy cattle	30%
Twinning in sheep	10%
Weaning weight in sheep	25%
Jumping ability in horses	20%

factors, such as nutrition, air quality, number of breedings per sow, or disease. In contrast, the heritability of carcass traits, such as ribeye area in beef cattle, is 50 to 60 percent. Thus, only 40 to 50 percent of the observed differences in ribeye area are a result of environmental influences. Growth and feed efficiency traits are intermediate in heritability, averaging 25 to 40 percent.

Traits that have high heritability are most responsive to genetic selection. For instance, mating a bull and a cow that are both heavily muscled will increase the chance that the offspring will also be heavily muscled. However, mating a boar and a sow from big litters does little to guarantee the offspring will also farrow big litters.

Interestingly, low-heritability traits show high levels of heterosis, while high-heritability traits show low levels of heterosis. In our example, we would expect less heterosis in a crossbred animal for ribeye size (a high-heritability trait) than for pigs born per litter (a low-heritability trait).

SELECTION

Using genetics to improve economically important traits in livestock has long been the objective of producers. The first rule in selective breeding is: Mate the best to the best. Identifying the best animals for a certain trait or set of traits requires good production records. For example, if a dairy farmer wants to keep replacement heifers from the highest-producing cows, the farmer must have records of which cows give the most milk. Likewise, if a beef producer is selecting for growth rate, the producer must keep weight records of possible replacement heifers and prospective herd bulls to know which ones gained weight the fastest. Producers then make selection decisions based on contemporary groups. A fall-born beef calf cannot be directly compared to a calf born in April because they were raised under different environmental conditions.

Selection intensity is the percentage of available animals from the top of the contemporary group that are used to produce the next generation. A population of animals will show a normal distribution for any

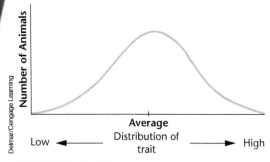

FIGURE 6-13

Normal distribution for any quantitative trait.

quantitative trait. See figure 6-13. The fewer the number of top-performing individuals that can be used as parents for the next generation, the faster the genetic improvement for that trait will be. In this example of selection intensity, 100 bulls were fed together in a contemporary group. Average daily gain was calculated for each. If the five bulls with the highest average daily gain were kept to sire the next generation, the selection intensity for average daily gain would be 5 percent. Livestock producers use artificial insemination (AI) to increase selection intensity for sires. Semen can be collected from proven top sires, diluted, and distributed to all parts of the country. Therefore, all producers have a chance to sample some of the very best genes available for a given trait. The selection intensity using AI may be less than 0.5 percent for sires. Because males can sire offspring by many females, the selection intensity for males is usually higher than that for females. Superovulation coupled with embryo transfer is one way to increase selection intensity in females.

Figure 6-14 illustrates the genetic change for a highly heritable trait using relatively high selection intensities for both males and females. Notice that the average trait value of each generation increases quickly with each generation. Low-heritability traits would respond similarly, but by a lesser degree per generation. Figure 6-15 shows the genetic change over three

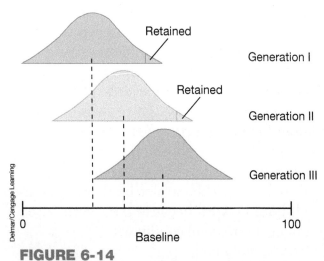

FIGURE 6-14

Change in baseline average for a highly heritable quantitative trait over three generations of high selection intensity.

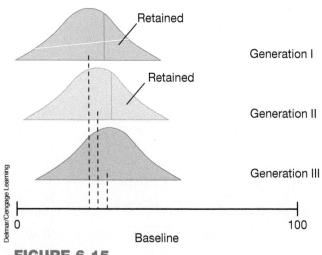

FIGURE 6-15

Change in baseline average for a highly heritable quantitative trait over three generations of low selection intensity.

generations using low selection intensities for both males and females.

Obviously, making genetic change relies on replacing the old generation with a new, genetically better generation. Therefore, the faster generations can be turned over, the faster genetic change can be made. The amount of time taken to replace each generation is called the **generation interval**. Generation interval for cattle is about five years, swine one year, sheep one year, and horses about four to five years. Species with shorter generation intervals can make more rapid genetic changes.

Selection for a single trait will cause faster improvement than selecting for many traits at the same time. For instance, if a sheep producer selects replacements based solely on wool yield, wool weights will increase relatively quickly, but fertility, longevity, and structural soundness may suffer. However, a swine producer selecting replacements based on three traits (backfat, feed efficiency, and pigs born per litter) may not see great improvement in any one area for several generations.

Courtesy of B.K. Lilija, Fort Valley State University

FIGURE 6-16

Blood samples can be taken either from the tail or the neck of cattle.

GENOMICS

Genomics is study of large portions of the entire genetic sequence of an animal with specific attention to portions that control economically important, quantitative traits. You will learn more about the practical applications of genomics in Chapter 14. Further research and understanding of the genome may shorten the generation interval by allowing producers to accurately select animals with favorable genes at birth by taking a simple blood sample, and then analyzing the DNA for combinations of favorable genes. See figure 6-16.

SUMMARY

Genes from both parents are mixed at fertilization. As a result of genetic mixing, offspring display traits from both parents. Genotype is the actual genetic code of the animal, while phenotype is that part of the genetic code that is expressed. Qualitative traits like coat color are controlled by a single set of genes. Quantitative traits, such as average daily gain, are controlled by many sets of genes and are economically important to producers. Inbreeding and line breeding are breeding systems where related animals are mated. This serves to concentrate genes from a common ancestor. However, crossbreeding results in heterosis, an improvement in performance over the average of the parents. Heritability is the proportion of a quantitative trait that is controlled by genetics rather than environment. Selection intensity and generation interval must be used in conjunction to realize rapid genetic change. Single-trait selection results in more rapid genetic change than multiple-trait selection.

CHAPTER REVIEW

EXPERIENTIAL LEARNING OPPORTUNITIES

1. Many land-grant universities or state-sponsored groups feed contemporary livestock groups for performance-testing purposes and then sell the animals to producers. Obtain the results from a performance test. (County extension educators or agriculture teachers can help secure the data.) Differentiate between quantitative and qualitative traits. Categorize each trait as high, medium, or low heritability. Do a poster display for a science project that compares and contrasts the performance of these animals.

2. Request copies of livestock or dairy sale catalogs. Identify any genomic test results listed in the sale catalogs. Research the specific genetic tests you find.

DEFINE ANY TEN KEY TERMS

alleles

codominance

complete dominance

contemporary group

crossbreeding

generation interval

genetics

genomics

genotype

heritability

heterosis

heterozygous

homozygous

inbreeding

incomplete dominance

line breeding

out breeding

phenotype
    qualitative traits

quantitative traits

selection intensity

QUESTIONS AND PROBLEMS FOR DISCUSSION

1. Name two scientists credited with the early development of the study of genetics.

2. Explain why males determine the gender of offspring.

3. Write the chromosome pairings for males and females.

4. Differentiate between genotype and phenotype.

5. Which can be influenced by environment, genotype or phenotype?

6. Give a specific example of a qualitative trait.

7. Give a specific example of a quantitative trait.

8. Name one breed of beef cattle that exhibits incomplete dominance in coat color.

9. Do roan Shorthorn cattle show codominance in coat color?

10. Animals with two copies of the same gene for a certain trait are said to be _____.

11. Write a heterozygous gene code for a black Angus cow.

12. Discuss the importance of feeding and managing all animals alike in a contemporary group.

13. Write one disadvantage of inbreeding.

14. Describe the difference between inbreeding and line breeding.

15. The main advantage of out breeding is _____.

16. The increase in performance above the parents' average is _____.

17. Give one example of a low-, a medium-, and a high-heritability trait.

18. Explain how heritability and heterosis are related.

19. What phrase is used to indicate the percentage of animals from the top of a contemporary group that are used as parents for the next generation?

20. What is the generation interval for cattle?

Ethology: Animal Behavior and Welfare

MONKEY SEE, MONKEY DO

Objectives

After completing this chapter, students should be able to:

▶ Give specific examples of maintenance, social, and learned behaviors in animals

▶ Identify behavioral traits that will respond to selection

▶ Apply the principles of animal behavior to animal movement and facilities

▶ Debate the difference between animal welfare and animal rights

Key Terms

animal rightists

animal welfarists

conditioning

eliminative

ethology

Flehmen response

flight zone

ingestive

learned behaviors

lip prehenders

maintenance behaviors

negative reinforcement

positive reinforcement

social behaviors

thermoregulatory behaviors

tongue prehenders

103

Career Focus

Kendra loved to watch her market hogs. Each year, from the time she was 8 until she went to college, she spent many hours each summer just watching her pigs. During this time she learned that pigs have specific behaviors: they always kept one corner of the pen as a bathroom and loved to play with any new object in the pen. She learned that pigs are very curious and eagerly investigated the family cat when it waltzed through the pen. The pigs' antics often made Kendra laugh out loud! In college, Kendra took a course in animal training and was exposed to the fascinating field of ethology. The course combined with her pig experience led her to a career as a zoo-animal behaviorist with the task of enriching the living space of captive animals from around the world.

INTRODUCTION

The science of the relationship between animal behavior and the environment has existed for ages, but it was not recognized as a legitimate field of study, called **ethology**, until relatively recently. Farmers in the dark ages undoubtedly noticed how animals acted when they were sick or in need of nourishment. More recently, animal behavior has been studied in an effort to better understand the physical needs of animals.

ANIMAL BEHAVIORS

Animal behaviors can be divided into three broad categories. The first is **maintenance behaviors**, such as eating, defecating, and sleeping. The second category is **social behaviors**, which deal with the interactions of two or more animals. The third category is **learned behaviors**. A horse adapting to carry a saddle and rider is an example of a learned behavior.

The study of animal behavior gives producers an understanding of how to design facilities to make moving and training easier for both livestock and their handlers.

Delmar/Cengage Learning

FIGURE 7-1

Knowledge of animal behavior ensures trouble-free livestock handling.

INDIVIDUAL MAINTENANCE BEHAVIORS

Individual maintenance behaviors include those behaviors basic to the sustenance of life. This category includes **ingestive** (eating), **eliminative** (defecating), and sleeping behaviors. Some animals also exhibit certain behaviors related to comfort, such as **thermoregulatory behaviors** related to maintaining an ideal temperature.

Ingestive behaviors vary among species. Ruminants (cattle and sheep) do not have incisor teeth on their upper jaw with which to bite off their food. Cattle grasp standing forages with their tongue and upper palate. See figure 7-2. Cattle are therefore known as **tongue prehenders**. Sheep are known as **lip prehenders** because they grasp forages with their lips. See figure 7-3. Sheep tend to nibble their feed and can graze closer to ground level than cattle. Horses and swine have incisor teeth on both jaws and can bite off chunks of feed like humans do.

All livestock species except swine excrete feces at random. Swine are particularly sensitive to where they defecate. Pigs are clean animals by nature and usually choose a corner of the pen to use for dunging. There are patterns, however, to exactly where this dunging occurs. For instance, pigs are more likely to defecate in an area that is the most draft-prone. Areas where pigs have nose-to-nose contact with pigs from another pen are also attractive dunging areas.

Sleep or deep rest is common in all livestock species. Pigs, like humans, sleep for extended periods. Judging from the jerking of pigs' legs during sleep, it appears that pigs dream. Sheep and cattle sleep lying down on their sides, but their sleep is sporadic and very short. Adult horses sleep in the standing position. See figure 7-4. A system of ligaments allows the horse to sleep while resting its muscles. While sleeping, horses rest their weight on only one hind leg with the other cocked. During sleep, horses switch the weight-bearing leg.

Pigs, in particular, display very obvious thermoregulatory behaviors. From simple observation a stockperson can tell if pigs are at a comfortable temperature.

FIGURE 7-2

Cattle grasp forages with their tongues and are known as tongue prehenders.

FIGURE 7-3

Sheep nibble forages with their lips and are known as lip prehenders.

FIGURE 7-4

Horses can sleep standing up.

FIGURE 7-5

Pigs display thermoregulatory behavior by piling on top of each other when they are cold.

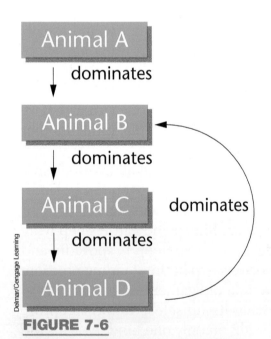

FIGURE 7-6

Animal dominance can be linear when each animal dominates a lower-ranking animal (A over B, B over C, C over D), or inverted when a lower-ranking animal rises to dominate a higher-ranking animal not directly above it (A over B, B over C, C over D, D over B).

If pigs are cold, they will lie together in a pile to take advantage of each other's body heat generation, and often lie with their legs pulled up underneath their bodies. See figure 7-5. Pigs that are too hot will lie so that one pig rarely touches another, and will lie flat on their sides. Comfortable pigs may lie touching one another, but will not be huddled together.

SOCIAL BEHAVIORS

Social behaviors of animals are very interesting and informative to those managing livestock. Behaviors that determine social structure or hierarchy among groups of animals have been studied extensively. In most animal groups, a pecking order is established. There is a boss animal that has asserted itself through bullying or fighting. The hierarchy of a group may not be completely linear. In a group of animals, Animal A may be dominant over B, B over C, and C over D. However, sometimes there can be an inversion where D is dominant over B. See figure 7-6.

When housed in small groups, swine have a nearly complete hierarchical social structure. In horses, groups of mares normally contain a dominant

Animal Science Facts

Researchers have spent hours patterning the eating behavior in pigs. Videos of various pigs' eating behaviors have been analyzed to determine the most common motion of a pig's head and mouth as it eats. This information was used to design a sow feeder that almost eliminates spilled feed. The feeder is J-shaped with the short curve of the J facing the sow. Several bars are installed horizontally inside the feeder, spaced along the long side of the J. When a sow takes a bite of feed from the bottom of the feeder, she naturally lifts her head and bumps her nose on the horizontal bars knocking the loose feed in her mouth back into the feeder. The feeder is deep enough so that, as the sow chews, the feed that spills from her mouth also falls back into the trough. This feeder, designed using the sows eating behavior as a guide, reduces feed wastage by more than 50 percent.

mare, but if a mature stallion is included in the group, he usually dominates the mares. Cattle dominance is based on age, with older animals in a herd usually more dominant over younger animals. Social dominance does not seem to be a component in the flocking behavior of sheep. See figure 7-7. Sheep simply want to be members of a flock. They do not seem to care who is boss.

Sexual behaviors are another example of social behaviors. Knowledge of sexual behaviors is extremely important to producers using artificial insemination, because they must recognize behaviors associated with estrus. Sows nearing estrus are restless, urinate frequently, and attempt to mount other sows. Cows exhibit the same signs. During heat, both sows and cows will stand to be mounted. See figure 7-8. Ewes show few signs of estrus except for the acceptance of the male. Mares in heat also urinate frequently, seek the companionship of other horses, and may lift their tail sideways.

In natural mating, sexual behaviors include both the male and female component. The male component in all species involves courtship. In swine, the boar will root at the flank and grunt aggressively before attempting to mount the sow. A ram's courtship patterns consist of walking with a stiff-legged gait and biting

© Pichugin Dmitry. Used under license from Shutterstock.com

FIGURE 7-7

Sheep tend to flock together and have little regard for social hierarchy.

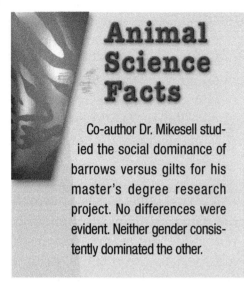

Animal Science Facts

Co-author Dr. Mikesell studied the social dominance of barrows versus gilts for his master's degree research project. No differences were evident. Neither gender consistently dominated the other.

Delmar/Cengage Learning

FIGURE 7-8

A cow standing to be mounted is a positive sign of heat.

the wool on the ewe's flank. The ram may also smell the urine of the ewe and curl his upper lip in what is called the **Flehmen response**. Bulls often smell or lick the cow's genitalia then exhibit the Flehmen response. Bulls will follow a receptive cow around the pasture, head resting on her rump. The stallion will smell the genitalia of the mare and may also exhibit the Flehmen response. In addition, a stallion may nip the mare with his teeth.

Social behaviors also include the behaviors exhibited between mother and young. Sows aggressively protect their newborn piglets against intruders. When a sow thinks her piglets may be threatened, she charges the intruder with mouth open. One reason farrowing crates are used by swine producers is to allow piglets to be handled without fear of confrontations with the sow. Baby calves are often born in open areas, such as pastures. A mother cow separates herself from other cattle as much as possible before calving. See figure 7-9. After the calf is born, the cow licks the calf dry stimulating the calf to nurse. The cow then spends the next day or two forming a strong bond with her calf before she and the calf rejoin the herd. Cows and calves can locate each other by smell, sound, or sight. Ewes, also, stimulate their young immediately after

FIGURE 7-9

Beef cows separate from the herd to calve in isolation. They gradually return to the herd when the calf is several days old.

birth by licking them. The bond between ewes and their lambs is obvious. In a pasture of several hundred ewes and lambs, a ewe will nurse only her own lambs. Mares also exhibit a bond with their foals and defend them against intruders. Voice recognition seems to be the most important means of communication between mares and foals.

LEARNED BEHAVIORS

Animals learn by three methods. The first is through **positive reinforcement**. See figure 7-10. Positive reinforcement is giving the animal something good (a treat, attention, feed) for doing something correctly. The second is **negative reinforcement**. Negative reinforcement is punishing the animal for doing something wrong. Animals can be trained in both ways, but generally, positive reinforcement works best. A well-placed scratch, some special feed, or simply a pat and soothing words are examples of positive reinforcement. Animals can regard simple management practices as negative reinforcement. For instance, if every time cattle are run through a catch chute, they get prodded with vaccination needles, they come to regard the chute with fear. The needle acts as a negative reinforcement. Some positive reinforcement combined with the shots would make the cattle less fearful.

A third type of learned behavior is known as **conditioning**. Animals may become conditioned to associate a certain stimulus with a behavior. For example, cows entering a milking parlor know from past experience that milking time is near. See figure 7-11. They respond to the parlor stimulus with a conditioned response called milk let-down.

FIGURE 7-10

Animals can be halter trained using positive reinforcement.

FIGURE 7-11

Cows waiting to enter a milking parlor.

BEHAVIORAL GENETICS

Behaviors can also be inherited. Behaviors are classified as quantitative traits, but they are usually more difficult to measure objectively than something like average daily gain. Producers know that certain breeds of animals are easier to handle and manage than others. For instance, beef bulls are generally known to be calmer and less dangerous than dairy bulls. See figure 7-12. The natural gaits of horses are a good example of a behavior that is genetically controlled. Tennessee Walking Horse foals naturally assume their distinctive gait with no training, and thoroughbred foals gallop across a field with the ease of their parents.

The American Angus Association has recently released genetic data for docility. Producers can use the data to select sires whose calves will be even tempered. Since calm cattle have been proven to convert feed to gain more efficiently than wilder cattle, producers see a direct financial benefit from selecting for docility. In pigs, one behavior that has received research attention is the sitting behavior. This trait is important because sows that spend considerable time sitting tend to crush more baby piglets than those that do

Delmar/Cengage Learning

FIGURE 7-12

Bulls can be unpredictable.

not. Researchers measured the amount of time sows from different families spent sitting down. Results proved that sitting behavior is genetically controlled. Pigs from some families spent considerably more time sitting than those from other families. Producers can use this information to select replacement gilts. They can, therefore, expect their sows not to sit and crush so many piglets.

BEHAVIOR AND LIVESTOCK MOVEMENT

The movement of animals from place to place requires knowledge of animal behavior. Body position is the key to moving animals. Each animal has a **flight zone** surrounding it that, if invaded, causes the animal to move away. See figure 7-13. In extremely tame animals, the zone may be nearly nonexistent. However, in wild animals the zone may be extremely large. Most farm animals are semi-tame and will adhere to the following rules. The handler's body position within the flight zone determines the direction the animal will move. If the animal is approached to the rear of the point of the shoulder, it will most likely move forward, angling to the opposite side of the handler. If an animal is

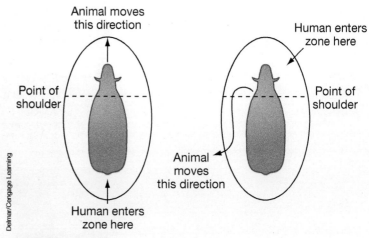

FIGURE 7-13

Animals move in the opposite direction when an approaching human enters the flight zone.

approached to the front of the point of the shoulder, it will most likely turn away from the handler and travel in the opposite direction.

Animals will balk at physical obstacles. Steps up or down or a change in flooring type will cause animals to stop and look. If animals are given enough time to investigate the steps or change in flooring, they will normally continue to move. Rushing animals will only worsen the situation.

Walls of livestock-movement alleys and loading facilities should also be modified to make livestock movement easier. Animals are easily distracted by other animals outside a passageway. Walls should be solid to avoid this problem. Alleyways should be either wide enough so that two animals can fit side by side, or narrow enough that only one animal can fit, without room to turn around. Sharp turns, especially blind turns should also be avoided. Turns should be curved. Cattle, especially, have a natural behavioral tendency to follow a curved passageway. See figure 7-14.

Air movement and light should also be considered when moving livestock. Strong drafts entering a doorway through which livestock are being moved can cause livestock to balk. Likewise, moving livestock

Delmar/Cengage Learning

FIGURE 7-14

Animals move easier around curved handling facilities.

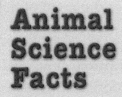

Animal Science Facts

Cow tipping is an old college hoax designed to trick naïve nonfarm students. The inexperienced students are told that cows sleep while standing up and that a group of determined people can push the cows over onto their sides. The tricksters then take the greenhorns to a cow pasture and tell them to give it a try. While the initiates are busy finding a receptive cow or attempting to tip an indignant bovine, they are gleefully abandoned. Don't fall for this old prank!

from bright light into darkness can cause movement problems, primarily because the animal perceives a dark shadow to be a bottomless hole into which it is reluctant to plunge. Movement from darkness to light does not seem to be as much of a problem. However, any obstacles should be eliminated before animal movement is attempted. If animals balk when moving through a handling system, it often helps if the handler moves through the passage at animal level to determine the cause of the problem.

Caution Statement: Care must always be taken around even seemingly tame animals. Too many fatalities have occurred when animals have turned on their handlers.

ANIMAL WELFARE AND ANIMAL RIGHTS

Many people mistakenly believe that animal welfare and animal rights are the same concept. Welfare is defined in the 1989 edition of *The New Lexicon Webster's Dictionary of the English Language* as being "healthy, happy, and free from want." The welfare of animals is, without question, the responsibility of the owner, producer, or caretaker. Most producers would consider themselves **animal welfarists**. On the other hand, those who advocate animal rights think that animals should have the same rights and privileges as humans. **Animal rightists** do not think animals should be used for any human benefit, such as food (milk, meat, eggs), clothing (leather, wool), or pleasure (horse racing, pleasure riding).

Animal welfarists contend that for animals to grow, reproduce, and perform at profitable levels, they must be well tended. Livestock under the care of competent producers have their basic needs met. These basic needs include feed, water, and protection from parasites, disease, and predators. Producers who are animal welfarists believe confining livestock in pens serves to keep the animals from many dangers, such as foul weather or busy roadways. Along that same line, confining sows to crates during gestation eliminates competition from other sows for feed. Housing sows in farrowing crates also reduces the number of baby piglets crushed by the sow, and increases the welfare of the piglets.

Animal rights advocates often have little connection with food and fiber production. However, they usually do have access to finances with which to influence lawmakers. Animal rightists want lawmakers to pass legislation making it illegal to raise animals for human use. They label contemporary livestock production as factory farming where livestock are routinely mistreated and abused. The best way for animal welfarists to avoid these types of laws is to support scientific study of animal behavior. Researching animal behavior is the best way for humans to objectively determine the needs and wants of animals.

SUMMARY

Maintenance behaviors are critical for life. Examples include eating, defecating, and sleeping. Social behaviors involve two or more animals and include hierarchy, sexual behaviors, and bonding between mother and young. Learned behaviors can be taught through both positive and negative reinforcement, or conditioning. Some behaviors are inherited as a quantitative trait. Animal movement can be eased through knowledge of animal behaviors and modification of facilities. Animal rightists assign the same privileges to animals that humans enjoy. Animal welfare involves making sure animals are happy, healthy, and free from want. Many management practices, although seemingly inhumane, actually increase animal welfare.

CHAPTER REVIEW

EXPERIENTIAL LEARNING OPPORTUNITIES

1. Many local theater companies contract animal behaviorists to train animals for shows. An example would be the trainer of the dog in the production of *Annie*. Contact a local theater company to connect with such an individual. Perhaps a class demonstration could be arranged.

2. Using an animal rights versus animal welfarist format, prepare a debate for the FFA agricultural issue forum.

DEFINE ANY TEN KEY TERMS

animal rightists

animal welfarists

conditioning

eliminative

ethology

Flehmen response

flight zone

ingestive

learned behaviors

lip prehenders

maintenance behaviors

negative reinforcement

positive reinforcement

social behaviors

thermoregulatory behaviors

tongue prehenders

QUESTIONS AND PROBLEMS FOR DISCUSSION

1. List three examples of maintenance behaviors.

2. List three examples of social behaviors.

3. List three methods of learned behaviors.

4. Give two advantages of farrowing sows in crates. Can you name any disadvantages?

5. Do ruminants have upper incisor teeth?

6. Which species, cattle or sheep, graze forages closer to the soil? Why?

7. True or False. Swine reserve a portion of the pen for defecation.

8. Which livestock species sleeps for extended periods of time in the prone position?

9. In which position do cattle sleep?

10. Social dominance in cattle is usually based on _____.

11. Is social dominance seen more in cattle or sheep?

12. True or False. Cattle prefer to give birth when their herd mates are nearby.

13. Which typically exhibit more dangerous behavior, beef bulls or dairy bulls?

14. Name three treats used in positive reinforcement of animals.

15. Milk let-down upon entry into a milking parlor is a learned behavior referred to as _____.

16. Explain how behavior and genetics can be related.

17. True or False. The sitting behavior in swine is linked to genetics.

18. Describe four physical obstacles that inhibit animal movement.

19. Cattle prefer to move in a _____ pattern.

20. Contrast the beliefs of an animal welfarist and an animal rightist.

2

Application

8

Swine Management and the Swine Industry

THIS LITTLE PIGGY WENT TO MARKET

Objectives

After completing this chapter, students should be able to:

▶ Give an overview of the swine industry in the United States

▶ Compare the physical characteristics of the eight major swine breeds and classify each as a maternal or paternal breed

▶ List and explain four breeding systems used in the swine industry

▶ Identify six steps in processing baby piglets

▶ Discuss nutrient requirements for various stages of swine production

▶ List and describe parasites and diseases common to breeding and market swine

▶ Discuss the housing requirements for swine

▶ Differentiate among cash, futures, and formula pricing

▶ Describe the organization of the swine industry

▶ Discuss two issues facing the swine industry

▶ List three careers in the swine industry

▶ Name two organizations that play supporting roles in the swine industry

Key Terms

all-in all-out production

atrophic rhinitis

backcrossing (crisscrossing)

biosecurity

cash price

check-off

cross-fostered

farrow to finish

formula price

futures price

independent producer

integrated production

leptospirosis

maternal breeds

mycoplasmal pneumonia

parvovirus

paternal breeds

PRRS

pseudorabies

rotaterminal crossbreeding

seedstock

split-sex feeding

swine erysipelas

terminal crossbreeding

three-breed rotational cross

wean to finish

Career Focus

Brent raised a couple of market hogs during his FFA years in high school, but he never really considered a career in production agriculture. Brent's real passion was education and he always considered himself an agriculture teacher in the making. However, after he completed college he couldn't find that perfect teaching job, so he took a temporary position as an assistant manager at an integrated sow farm. During his first few weeks on the farm, his superiors noted Brent's willingness to use his college education to explain to other employees the reasons for doing their many daily tasks. Management noticed that, because employees knew the reasoning for the many jobs they had to perform, employee morale improved significantly. As a result, productivity of the sow farm increased dramatically. Within six months, Brent was asked to begin a formal training program for new employees in the company's swine production system. Here, he used his passion for teaching combined with his college technical education in an industry setting.

INTRODUCTION

Historically, pigs have paid for many farms in the United States, earning their title as the mortgage lifter. The business of swine production has completed a massive industry consolidation. Most producers now function as part of an **integrated production** system where a few producers own and control a majority of the pigs as well as the feed mills and slaughter facilities servicing them. Few **independent producers** (not aligned with a larger business entity) remain. No matter how the industry changes, pork will still be a favorite meat of American consumers. Pigs will continue to be raised by American producers and the principles of swine production will remain the same. Importantly for students, the swine industry offers various job opportunities, ranging from herd managers to food scientists to advertising specialists.

INDUSTRY OVERVIEW

The trend toward fewer but bigger farming operations is perhaps most evident in the swine industry. In the early 1960s, over 1 million American farms raised an

FIGURE 8-1

Though the swine industry is largely integrated, students still enjoy cute piglets.

Courtesy of USDA. Photo by Scott Bauer

Animal Science Facts

Before the advent of the pickup truck or semi-trailer, hogs were driven on foot, sometimes for hundreds of miles, from the farm to major pork-packing cities. Cities that both earned the reputation as Porkopolis were Cincinnati and Chicago.

average of 50 head per year. In 2008, only 73,150 farms inventoried 67.4 million hogs. Eighty-five percent of the nation's hogs were housed on farms with inventories reaching over 2,000. Another 8 percent were on farms with numbers in the 1,000 to 2,000 head range. The remaining 7 percent of hogs are on farms with less than 1,000 head. These smaller farms account for 60.9 percent of all swine operations.

Traditionally, the swine industry was centered in the Corn Belt where feed is readily available. Since feed accounts for 65 percent of total swine production costs, and finishing pigs from 40 pounds to market weight consume a significant amount of feed, the Corn Belt remains the heart of hog-finishing operations. However, the past 30 years have seen sow operations move from the Corn Belt to states with a history of integrated poultry production, so lenders and farmers were familiar with the livestock integration model. North Carolina led the way in rapid swine production growth. In 2008, 99 percent of all North Carolina swine were located on facilities holding 1,000 or more hogs. Most of these large operations are highly integrated with owners controlling many aspects of production. Large North Carolina operations average over 2,500 head. Because of public opposition, cost of transporting corn from the mid-western states, and nutrient management issues, construction of new swine operations in North Carolina appears to have slowed.

(Data in the Industry Overview section obtained from the 2007 USDA Census of Agriculture)

BREEDS

The United States has eight traditional breeds of swine. Each of them has, at various times in history, held a place in crossbreeding schemes practiced by swine breeders. Some breeds excel in maternal traits, such as pigs born per litter or milking ability, and are called **maternal breeds** or dam breeds. **Paternal breeds** or sire breeds are those that excel in growth rate, muscling, or leanness. Since 1990, swine breeders and companies that specialize in swine genetics

have sought other breeds from around the world for either their maternal or their paternal traits. In this section, we will discuss the eight major swine breeds, as well as other, more recently imported breeds from around the world.

BERKSHIRE

Berkshires trace their ancestry to Berkshire, England, and were first imported to the United States in the 1820s. See figure 8-2. Berkshire hogs can be distinguished by their black color, six white points (four feet, nose, and tail), and upright ears. At one time, Berkshires were known for their short pug noses. However, since the early 1970s, breeders have concentrated on breeding pigs with longer noses. The longer nose facilitates the animal's ability to eat from automatic self-feeders. Berkshires can function as either a maternal or a paternal breed because they are not dominant in either trait. For this reason, they may be used in crossbreeding systems where both maternal and paternal traits are important. Berkshires are known to have exceptional muscle quality in terms of color, texture, and flavor. These traits result in exceptional meat palatability, so the popularity of Berkshires has mushroomed due to demand from high-end eating

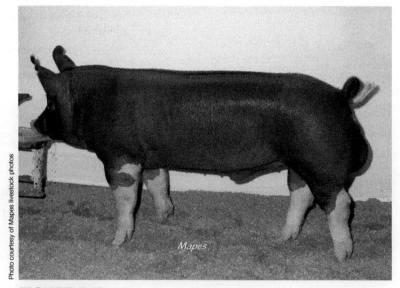

Photo courtesy of Mapes livestock photos

FIGURE 8-2

Berkshires are black with six white points and upright ears.

establishments and overseas importers. In 2008, 5,416 litters of Berkshires were recorded.

CHESTER WHITE

Chester Whites were developed in Chester County, Pennsylvania, from several other breeds of hogs in the early 1800s. See figure 8-3. Chester Whites are obviously white with rather small, droopy ears. This breed is known to be exceptional in mothering ability and would be classified as maternal. They are easily distinguished from Landrace by a generally increased muscularity, larger bone diameter, and much smaller ear size. Chester Whites from 5,666 litters were recorded in 2008.

DUROC

Durocs originated from a breed of hogs called Jersey Reds from New Jersey. See figure 8-4. These were crossed, over a period of years, with another red breed from New York called Durocs. The resulting breed established in the 1860s was originally called Duroc-Jerseys. They are now simply known as Durocs, which can have a variety of red shades with small, drooping ears. Durocs are best known for their growth rate and

Photo courtesy of Mapes livestock photos

FIGURE 8-3

Chester Whites are a white maternal breed with small, drooping ears.

Photo courtesy of the National Swine Registry

FIGURE 8-4

Durocs are a red paternal breed with drooping ears.

carcass traits and would be classified as a paternal breed. In the past, Durocs were exported to other countries where breeders spent many years selecting for extreme leanness and muscling. In the 1980s American breeders re-imported some of these extremely lean Durocs into the United States to improve carcass composition. Imported Durocs served to further improve American Durocs as a paternal breed. Most of the imports originated from Canada or Denmark and were called Canadian Durocs and Danish Durocs, respectively. Pork from Duroc-sired pigs is generally of higher quality than that of some other breeds. In 2008, 8,375 litters of Durocs were recorded.

HAMPSHIRE

Hampshires are black with a white belt circling the shoulders. See figure 8-5. They originated in Boone County, Kentucky, where the breed association was formed in 1893. Hamps, as they are known, are often used as both maternal and paternal breed. They are known for their muscling and leanness, but Hampshire crossbred females are just as likely to be used as commercial sows. Like Durocs, Hampshires were exported and have been re-imported, mainly from Sweden and Canada. The National Swine Registry recorded 4,451 litters in 2008.

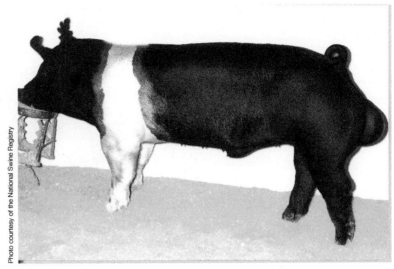

FIGURE 8-5

Hampshires are black, belted hogs that can function as either a paternal or maternal breed.

LANDRACE

The Landrace breed was first imported from Denmark in the 1930s. See figure 8-6. Since then, imports have arrived from both Norway and Sweden. These pigs are white and extremely long bodied with large, drooping ears. They are used almost exclusively as a maternal breed because of their large litters and their exceptional milking ability. In 2008, 3,328 Landrace litters were recorded.

FIGURE 8-6

Landrace are a white maternal breed with large drooping ears.

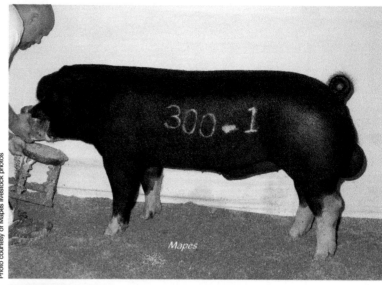

Photo courtesy of Mapes livestock photos

FIGURE 8-7

Poland Chinas are a black paternal breed with drooping ears and white on the nose, tail, and feet.

POLAND CHINA

Poland Chinas, or Polands, were first bred in Butler and Warren counties of Southwest Ohio, in the late 1800s. See figure 8-7. Their appearance is black with white points, much the same as Berkshires, except Polands have drooping ears. Polands are known for their growth rate, muscling, and hardiness. Therefore, their main use is as a paternal breed. A total of 1,067 Poland litters were registered in 2008.

SPOTTED SWINE

Spotted Swine, or Spots, were developed in the state of Indiana from predominantly the same base stock as Polands. See figure 8-8. The Spotted Swine Association was formed in 1914. As indicated by the name, Spotted Swine have black and white spots and drooping ears. Spots are also used as a paternal breed in crossbreeding systems. Registrations of Spotted Swine litters numbered 2,376 in 2008.

YORKSHIRE

Yorkshires, or Yorks, are a maternal breed of great renown. See figure 8-9. White with upright ears,

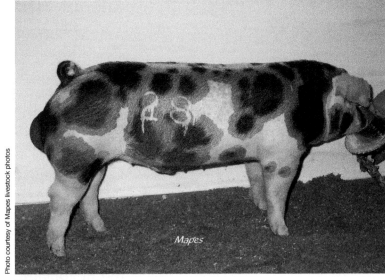

FIGURE 8-8
Spotted Swine are black and white and serve as a paternal breed.

FIGURE 8-9
Yorkshires are a white maternal breed with upright ears.

they play an important role in many crossbreeding systems because of their ability to farrow and wean large, heavy litters. Yorks originated in and around the county of Yorkshire, England, and were imported in the early 1800s. European Yorkshires are known as Large Whites. In the 1980s Large Whites were imported in an attempt to improve American Yorkshires. Yorkshires are the most popular purebred swine with registrations from 12,900 litters in 2008.

OTHER BREEDS

Another breed less commonly used by commercial swine breeders is the Tamworth, a red breed with upright ears. See figure 8-10. Tamworths were originally developed in Ireland and imported to the U.S. in the early 1800s. Tamworths are popular with free-range, organic producers. Tamworth meat is said to be especially flavorful. Herefords (red with white face, legs, and underline) are another relatively minor breed, developed in the United States in the early 1900s.

A more recent addition to America's swine breeds is the Pietrain from Germany and Belgium. See figure 8-11. Most Pietrains are black and white spotted in appearance and are generally extremely lean and heavily muscled. Many Pietrains carry two copies of the stress gene, which adversely affects muscle quality. PSE (pale, soft, and exudative) muscle is common from Pietrain and Pietrain crossbred pigs. Their main use is to be crossed with other paternal breeds to produce extremely lean, heavily muscled, crossbred terminal sires with zero or one copy of the stress gene. Pietrain use peaked in the early 1990s and has since declined because, even though meat yield is exceedingly high in the carcass, Pietrain meat quality is relatively poor. There is no breed association for Pietrains in the United States.

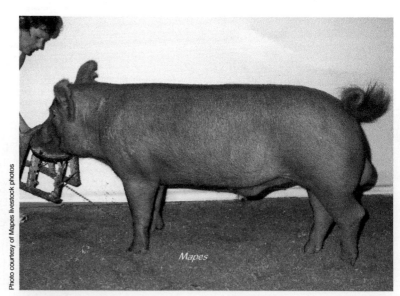

Photo courtesy of Mapes livestock photos

FIGURE 8-10

Tamworths are a red breed with upright ears and are known for their meat quality.

Chinese strains of swine were imported in 1989 by American universities to study their amazing maternal abilities. See figure 8-12. Chinese breeds routinely farrow 15 to 20 pigs per litter but are extremely fat and light-muscled by U.S. standards. The goal of researchers was to isolate the genes responsible for these large litters and transfer them to leaner, heavier-muscled breeds.

(Number of animals registered for individual swine breeds courtesy of the National Swine Registry and Team Purebred.)

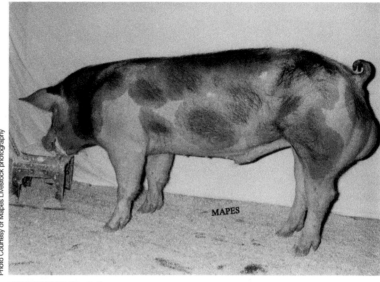

Photo Courtesy of Mapes Livestock photography

FIGURE 8-11

Pietrains were bred to be extremely lean and heavily muscled.

Photo courtesy of USDA

FIGURE 8-12

Swine scientists marvel at the amazing maternal abilities of Chinese pigs.

BREEDING SYSTEMS

Producers have many options for crossbreeding pigs to maximize heterosis and to best incorporate the relative maternal and paternal advantages of each breed. Producers traditionally used the previously described purebred hogs in these crossbreeding systems, but most commercial swine genetics companies now use synthetic (inbred, human-made) breeds. Discussed here are four crossbreeding systems.

BACKCROSSING

The first crossbreeding system is called **backcrossing** or crisscrossing. See figure 8-13. In a backcrossing system, only two breeds are used—usually one maternal and one paternal. At the beginning of this system, breed A is mated to breed B. Replacement gilts are kept and bred to another boar of breed A. Replacement gilts from the third generation are retained and mated back to a boar of breed B and so on. This system does little to maximize heterosis after the first generation because the dam will always contain some of the same genes as the boar to which she is bred.

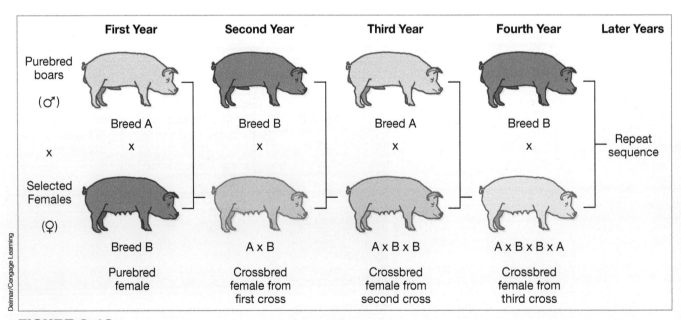

FIGURE 8-13

A backcrossing or crisscrossing system involves just two breeds.

THREE-BREED ROTATIONAL CROSSBREEDING

The second crossbreeding system is called a **three-breed rotational cross**. See figure 8-14. Three-breed systems normally contain one maternal, one paternal, and one combination breed. This system is similar to backcrossing, except a third breed is introduced into the program. Breeds A and C are crossed; retained gilts from the crossbred A and C litters are mated to boars of breed B. Gilts from this generation are mated back to breed boars of A and so on. The three-breed rotational cross maximizes heterosis better than the backcrossing system. However, after three generations, dams contain some of the same genes (although fewer than in the backcrossing system) as the boars to which they are bred.

ROTATERMINAL CROSSBREEDING

The third system is called a **rotaterminal crossbreeding** system. See figure 8-15. Three maternal breeds are required for this system. These three maternal breeds are treated like a three-breed rotational cross. The three-way cross gilts from this rotational program are bred to paternal boars that are extremely lean and

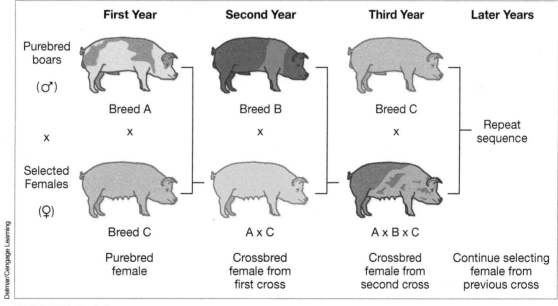

FIGURE 8-14

A three-breed rotational crossbreeding system adds a third breed to the crisscrossing system.

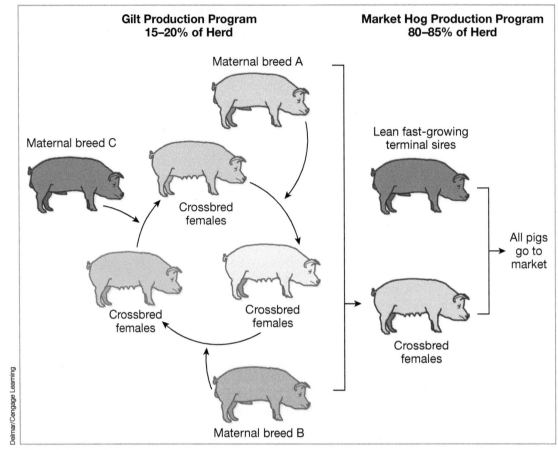

Delmar/Cengage Learning

FIGURE 8-15

The rotaterminal crossbreeding system utilizes three maternal breeds and a terminal sire breed.

heavily muscled. These boars are referred to as "terminal sires" because no gilts from their litters are kept for replacements. All the resulting pigs are destined for slaughter. This system maximizes heterosis in the terminal pigs headed for slaughter, but not in the sow herd. A disadvantage of this system is that it requires the maintenance of at least four breeds of boars—a difficult task for small producers.

TERMINAL CROSSBREEDING

The fourth and most effective system for maximizing heterosis is a **terminal crossbreeding** program. See figure 8-16. In a terminal program, producers buy crossbred gilts of two or more maternal breeds from reputable producers or breeding stock companies. Large farms reserve a portion of the sow herd for creating these females. These crossbred

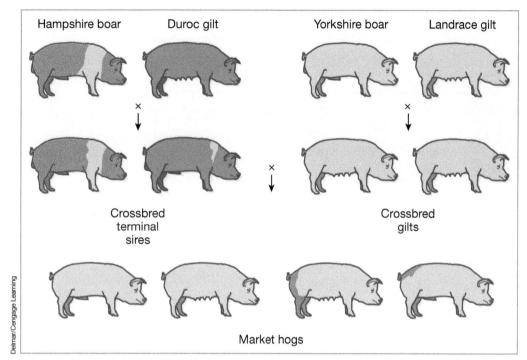

Market hogs

Delmar/Cengage Learning

FIGURE 8-16

The terminal crossbreeding system maximizes heterosis by using a planned crossbreeding system where no breeds are used twice. All resulting pigs are destined for slaughter.

gilts with 100 percent heterosis are mated to terminal sire boars, which may also be crossbred, but are extremely thick, heavily muscled, and selected for outstanding pork quality. The resulting pigs are strictly destined for slaughter and no gilts are retained. The vast majority of commercial market hogs in the United States are a result of a terminal crossbreeding program.

SWINE PRODUCTION CYCLE

Gilts destined to enter the breeding herd are typically mated at about eight months of age. Pregnant gilts and sows spend the 114-day gestation period either in gestation stalls or penned with groups of contemporaries. Gestation stalls allow producers to ensure that each individual can be fed and managed to its needs. With the advent of computerized feeding systems and electronic ear tags that allow grouped animals to be fed as individuals, larger producers are gradually switching to open pen gestation systems.

Animal Science Facts

Swine Facts

Average body temperature—102.5 degrees F

Average pigs weaned per litter (2008)—9.6

Average live market weight—270 to 275 pounds

Average 10th rib backfat—0.8 inches

Average loin eye area—7.0 square inches

Average daily feed intake (feed), feed per pound of gain (F/G), and average daily gain (ADG)

	Feed (lb)	F/G (lb)	ADG (lb)
40 lb pig	2.8	2.2	1.3
140 lb pig	4.6	2.9	1.6
240 lb pig	7.0	3.7	2.0
Avg. 40 to 240 pounds	5.0	2.8	1.8

Near the end of the gestation period, pregnant females are moved to a farrowing stall, which is a special pen equipped with space around the exterior so the piglets can escape from the mother. See figure 8-17. Mother sows can easily crush their newborn three to four pound piglets, and farrowing stalls help to prevent crushing losses. After the sow farrows, she typically nurses the litter for about 18 to 21 days prior to weaning.

At weaning, the sow is removed from the farrowing stall and returned to gestation quarters. Typically three to seven days after weaning, the sow will return to heat and is remated for her next litter. If she becomes pregnant during her first heat cycle, she will farrow again 114 days later. If not, she should come in heat again 21 days later. One of the production numbers producers watch with care to determine efficiency in a swine breeding herd is litters per sow per year (L/S/Y). Biologically, a sow is limited to one farrowing every 140 days (114 days of gestation plus 21 days of

Photo courtesy of Chore-Time Hog Production Systems, Milford, IN

FIGURE 8-17

Farrowing stalls are designed to prevent the sow from crushing her piglets.

lactation plus 5 days to rebreed). At this rate a single sow can have as many as 2.6 litters per year (365/140). This number, averaged over an entire herd, is one indication of the herd's reproductive performance.

At the end of their productive life (typically about six litters), sows that fail to rebreed, have structural deficiencies, or routinely farrow small litters are culled. Meat from most culled sows ends up as sausage.

PIGLET AND NURSERY MANAGEMENT

Baby pigs have a very limited ability to control their own body temperature. The temperature requirement for newly born pigs is 90 to 95 degrees F. However, sow feed intake declines when the temperature gets much above 70 degrees F. Therefore, the room temperature should be set where the sow is comfortable, and supplemental heat via a heat lamp or heated mat should be provided for piglet comfort. Supplemental heat can be gradually decreased to 75 to 80 degrees F at three to four weeks of age. Most commercial sows give birth in farrowing stalls over slotted floors where manure and urine can easily drain away. Smaller producers without the luxury of slotted floors often use bedding for sows and baby pigs. The bedding serves to soak up waste while substantially helping baby pigs to stay warm.

When piglets are born, they should receive colostrum, or sow's first milk as quickly as possible. Antibodies present in colostrum help protect piglets from disease for the first three weeks of life. Within a day or so of birth, piglets will claim one of the sow's 10 to 16 nipples as its own and will only nurse from that nipple until weaning.

Newborn piglets require some special management procedures within the first few days of life. These procedures are collectively called baby pig processing.

If the farrowing is attended, each piglet's umbilical cord should be cut about four inches from the navel shortly after birth. The remaining cord should

then be dipped in or sprayed with iodine. The next step to processing baby pigs is to cut needle teeth. Piglets are born with eight sharp teeth, four each on the upper and lower jaws. The tips of these teeth can be removed with small, side-cutting pliers when the pig is one day of age. See figure 8-18. If not removed, the needle teeth can easily slice the noses of other piglets during fights over teats. Piglets can also bite the sow's udder making it painful for sows to nurse. Open wounds could lead to infection or the sow's refusal to nurse. However, some large commercial swine operations have abandoned this traditional practice and experience few of these problems.

Piglets' tails are often docked to half their original length at the same time as the needle teeth are cut. See figure 8-19. Older pigs, if fed in confinement conditions, sometimes bite other pigs' tails. Docking tails at a young age reduces the chance of later tail biting. Ear notching for identification of individual pigs can also be performed at the same time as teeth clipping and tail docking. See figure 8-20. The notches in the left ear indicate the pig's number within the litter and notches in the right ear indicate the litter number. Traditionally, producers use number one for the first litter born after January 1. Subsequent

Photo by Frank Flanders

FIGURE 8-18

Piglets' teeth may be clipped to prevent injuring the sow's udder.

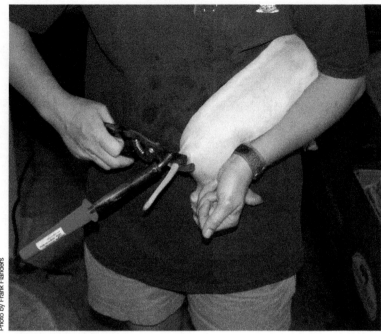

Photo by Frank Flanders

FIGURE 8-19
Piglets' tails are docked to prevent potential tail biting later in life.

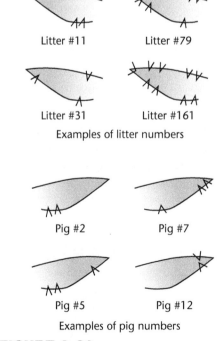

Key to standard ear notching system

Litter #11   Litter #79

Litter #31   Litter #161

Examples of litter numbers

Pig #2   Pig #7

Pig #5   Pig #12

Examples of pig numbers

Courtesy of Indiana Cooperative Extension Service, Purdue University

FIGURE 8-20
Ear notching is one way to identify pigs.

litters are numbered two, three, and so on. Large swine operations have adopted their own notching systems. The notching system is more permanent than ear tags that may be lost. Pigs may also be identified using tattoos.

Piglets should be given an appropriate dose of long-lasting antibiotic one day after their birth to help ward off infections. Sow's milk is low in iron; therefore, to prevent anemia, piglets should be given injectable iron at about seven days of age. See figure 8-21. Boars may be castrated at three to seven days of age to reduce the stress of this procedure. All injections should be given on the side of the neck about one inch behind the ear to avoid potential abscesses in the ham muscle.

If a group of sows is farrowing about the same time, piglets from one sow, which has too many piglets or too few functioning nipples, can be **cross-fostered** (transferred) to other sows with fewer piglets or more functioning nipples. Along the same line, smaller, weaker piglets can be cross-fostered onto heavier milking sows. This cross-fostering gives the weaker piglets a better chance to catch up with the average size of the group.

In most large swine production facilities, piglets are weaned at 21 days of age or about 12 pounds.

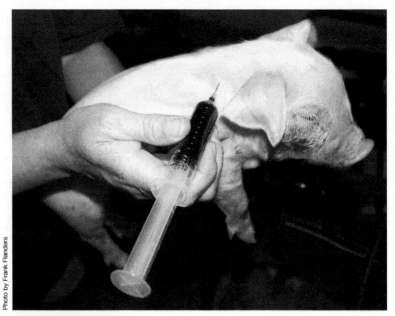

Photo by Frank Flanders

FIGURE 8-21

Baby pigs receive an iron injection to prevent anemia.

Ideally, they should be weaned by size, with big and small piglets (runts) placed in separate pens. In some cases, small pigs may be transferred to a nurse sow for several additional days of nursing after their contemporaries are weaned. In other cases, small piglets may be humanely euthanized. Piglets should be weaned into a cleaned, disinfected nursery room. See figure 8-22. Piglets usually remain in the nursery until they are nine to ten weeks of age and weigh about 50 pounds, when they are moved to a cleaned, disinfected finishing facility. Alternately, 12-pound piglets may move directly to a cleaned, disinfected finishing farm in a **wean to finish** system. The strategy of keeping pigs in separate rooms or buildings according to age, is called segregated production. The fact that all pigs are removed from a building or room before the next group of pigs arrives is called **all-in all-out production**. Wean to finish and all-in all-out production both serves to prevent or reduce the spread of disease among pigs.

© Emily Veinglory, Used under license from Shutterstock.com

FIGURE 8-22

Piglets are moved to a clean, disinfected facility as a part of all-in, all-out management.

SWINE NUTRITION

Feed requirements differ depending on the sizes and ages of pigs. For example, the nutrient requirements of mature, pregnant sows are drastically different from those of nursery pigs. Diets can be classified as follows: nursery, grower, developer, finisher, gestation, and lactation.

Nursery pigs, especially those weaned at 21 days of age, are perhaps the most challenging age group to feed and manage properly. This is because the pigs' digestive system is not fully developed and cannot digest grains, such as corn and soybean meal, that are routinely fed to older pigs. Young pigs are used to drinking sow's milk and often need a few days to adjust to eating solid feed of any kind. For this reason, nursery feeds usually contain some milk-based feedstuffs along with plant-based feeds that are easily digested. Figure 8-23 shows a formula of a basic nursery feed for 21-day-old pigs. Nursery feeds should be high in protein and energy to meet the requirements of young, rapidly growing pigs. As pigs get older and their digestive system becomes more mature, the expensive, milk-based feedstuffs gradually are replaced by less-expensive grains. Most

	Nursery Diet	Grower Diet	Developer Diet	Finisher Diet	Gestation Diet	Lactation Diet
Shelled corn or milo	820 lbs	1440 lbs	1540 lbs	1640 lbs	1100 lbs	1455 lbs
Soybean meal (48%)*	500 lbs	500 lbs	400 lbs	300 lbs	200 lbs	375 lbs
Rolled oat groats	200 lbs					
Dried skim milk	200 lbs					
Dried whey	200 lbs					
Dried fat	20 lbs					100 lbs
Oats					620 lbs	
Vitamins, minerals, & salt**	60 lbs	60 lbs	60 lbs	60 lbs	80 lbs	70 lbs
	2000 lbs	2000 lbs	2000 lbs	2000 lbs	2000 lbs	2000 lbs
% Protein	21.7	18.5	16.5	14.6	13.5	16.1

*Soybean meal can be purchased in combination with salt, vitamins, and minerals. This mixture is called a protein supplement.

**Vitamins, minerals, and salt are sold as a vitamin/mineral premix.

Delmar/Cengage Learning

FIGURE 8-23

Formulas for typical nursery, grower, developer, finisher, gestation, and lactation diets.

nursery feeds contain some kind of antibiotic to help ward off diseases and infections during this stressful period. Nursery pigs convert feed to gain at about 1.5 pounds of feed per pound of gain.

Grower diets are normally fed to pigs weighing between 50 and 110 pounds. Although lower in protein than nursery diets, grower diets provide plenty of amino acids (because of increased feed consumption) during the period of rapid growth when pigs deposit large amounts of muscle protein.

As nursery pigs begin the finishing phase of production, they are often switched to a developer diet. Developer diets, which have a lower percentage of protein than grower diet, should be fed to pigs weighing between 110 and 180 pounds. Because the total feed consumption of pigs is increasing, the total amount of protein eaten can be maintained even with a feed containing a lower percentage of protein. When pigs weigh 180 pounds, protein requirements again drop. Pigs should be fed a finisher diet at this time. In some highly managed feeding programs, intermediate diets are fed to more closely match the dietary needs of pigs throughout the growing phase. See figure 8-23 for grower, developer, and finisher diets. Feed efficiency for pigs from grower to finisher is normally about three pounds of feed per pound of gain. Feed intake will increase from about three pounds per day for a 50-pound pig to about seven to eight pounds per day for a 250-pound pig.

Pigs may be fed differently according to sex. Since gilts inherently deposit more lean muscle tissue and less fat than barrows, their protein requirements are higher. Therefore, if gilts are penned separately from barrows in the finishing phase, producers can use higher-protein grower and developer diets for the gilts than for the barrows. This practice is called **split-sex feeding**.

The nutrient requirements of gestating sows are very low since there are few nutritional demands placed on these sows. The developing fetuses demand relatively few nutrients above those required for sow maintenance. Gestating sows are normally fed four to six pounds of feed per day of a low-energy, low-protein diet. See figure 8-24. However, the diet must

Photo courtesy of Kenneth Kephart

FIGURE 8-24

Gestating sows are fed individual amounts of feed based on stage of production and body condition.

be higher in vitamins and minerals than normal since the sow's intake is low. See figure 8-23 for a gestation diet that, if fed at four to six pounds per day, will maintain or slightly improve sow condition. Boars are normally fed four to six pounds of the gestation diet.

Lactating sows, however, have tremendous nutrient demands placed upon them by a large litter of nursing piglets. Lactating sows should be fed as much high-protein, high-energy feed as they will consume (usually 10 to 20 or more pounds per day). If lactating sows do not eat enough high-quality feed to maintain body weight, they may deplete body reserves and be too thin to rebreed. See figure 8-23 for a typical lactation diet.

SWINE PARASITES

There are many swine parasites, but those of major economic importance are limited to worms, mange, and lice. Confinement of pigs to indoor quarters has greatly reduced the economic loss associated with internal and external swine parasites.

WORMS

A heavy load of worms in the digestive tract can reduce growth rate and cause organ damage. See figure 8-25.

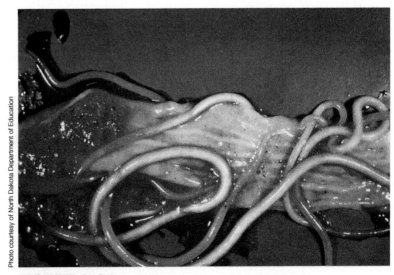

Photo courtesy of North Dakota Department of Education

FIGURE 8-25

Round worms reduce pig growth rate by robbing nutrients.

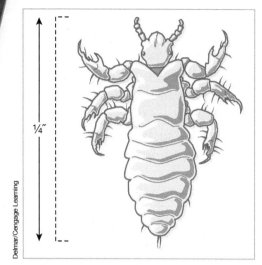

Delmar/Cengage Learning

FIGURE 8-26

Lice cause severe itching in pigs. They can be easily observed on infested areas of the skin.

A variety of worm species affect different organs, but all are easily controlled by feed or water treated with a commercially available wormer.

MANGE

Mange mites burrow under the skin and are a common problem in all areas of the United States. Scratching and hair loss are common signs of mange infestation. Even minor mange infestation will reduce growth rate and worsen feed efficiency. Mange is difficult to eradicate since mites can stay alive in bedding for several days. Adult mange mites can be controlled by insecticidal sprays, but re-treatment is often necessary since many products do not kill mange mite eggs.

LICE

Lice are small insects that spend their entire life on the skin of their host, in this case, the pig. See figure 8-26. Lice usually congregate around the ears and under the flanks where they cause severe itching. Like worms and mange, lice can reduce animal performance. Insecticidal sprays can be used to control lice. Several available broad-spectrum insecticides kill parasites of all types.

A parasite control program is essential to hog producers. Sows should be treated when moved to the farrowing room. This practice frees the sow of any parasites and reduces the chances of spreading them to the piglets. Piglets should be treated for parasites at about 50 pounds, and wormed again at about 130 pounds, if worms are a problem.

SWINE DISEASES

Many diseases affect swine to varying degrees throughout the United States. Entire books have been written on the subject. The more economically important diseases include atrophic rhinitis, erysipelas, leptospirosis, mycoplasmal pneumonia, parvovirus, PRRS, and pseudorabies.

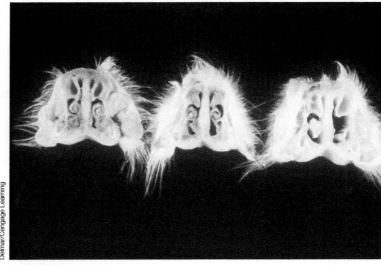

Delmar/Cengage Learning

FIGURE 8-27

Rhinitis causes the degeneration of the turbinate bones of the snout. The snout on the left is normal while the snout on the right is severely affected.

ATROPHIC RHINITIS

Atrophic rhinitis is an infection of the nose causing the degeneration of the turbinate bones inside the snout. See figure 8-27. The result is a twisted snout, which makes eating difficult. The degeneration of the turbinate bones (which function to filter incoming air) also allows bacteria easier access to the lungs. Aside from twisted snouts, sneezing is a primary sign. Rhinitis can be controlled using all-in all-out production in combination with vaccine administered to pregnant sows and baby pigs.

SWINE ERYSIPELAS

Classic **swine erysipelas** is readily diagnosed by the observation of raised diamond-shaped patches on the skin. Pigs may also have swollen ears, nose, and legs. Erysipelas is caused by a bacterium. Vaccination of gilts upon arrival and sows after weaning gives adequate protection. Typically, erysipelas cases are easily controlled by antibiotic treatment.

LEPTOSPIROSIS

Leptospirosis causes severe reproductive failure manifested by abortions or dead or weak pigs at birth.

Leptospirosis is easily prevented by vaccination of sows and gilts before breeding. All sows should be routinely vaccinated for leptospirosis as an insurance measure.

MYCOPLASMAL PNEUMONIA

Mycoplasmal pneumonia is a long-lasting disease that can exist with very few outward signs. Coughing in the morning is the most common sign indicating the presence of mycoplasmal pneumonia in the herd. Mycoplasma does cause reduced growth rate and poor feed conversion even though pigs do not look sick. Mycoplasmal pneumonia makes pigs more susceptible to other, more harmful respiratory diseases. If mycoplasmal pneumonia is diagnosed as a problem, vaccination of feeder pigs, combined with good management can successfully combat it.

PARVOVIRUS

Parvovirus is characterized by failure of sows to coming into heat, farrowing of small litters, and giving birth to mummified piglets. See figure 8-28. Prevention is easily accomplished by vaccinating all sows and gilts before breeding. Breeding females should be vaccinated for parvovirus as insurance whether or not disease signs have been observed.

Delmar/Cengage Learning

FIGURE 8-28

The piglet on the bottom of the photo is mummified.

PORCINE REPRODUCTIVE AND RESPIRATORY SYNDROME

PRRS (porcine reproductive and respiratory syndrome) is a viral disease that has two components affecting different stages of swine production. The first component is the reproductive component. Sows show irregular estrous cycles and have reduced conception rates, small litters, and small, weak piglets. The second component is the respiratory component that makes the respiratory system of the pigs extremely susceptible to other diseases. Normally, a herd will experience an outbreak of PRRS, and then develop an immunity that lasts one to three years. Subsequent outbreaks are usually less severe. The PRRS virus can be isolated from many swine herds. A PRRS vaccine that controls the disease with varying degrees of efficiency is available for baby pigs and sows. Regardless whether a herd has PRRS or not, isolation of incoming breeding stock is critical to keeping the virus from circulating within an existing breeding herd.

PSEUDORABIES

Pseudorabies (PRV) is caused by an easily spread virus and is usually most prevalent in areas where large numbers of hogs are raised. Symptoms include the death of large numbers of baby pigs less than three weeks of age. Abortion is also a common symptom. Some infected herds, however, may show few signs at all. Steps to control the spread of pseudorabies include 30-day isolation and blood testing of all new breeding stock. PRV has been eradicated (or nearly so) from many states. The practice of good **biosecurity** (methods to reduce disease transfer among pigs) helps control the spread of all swine diseases.

SWINE HOUSING

Pigs can be raised successfully under a wide range of conditions. Many pigs in the temperate climates are raised outdoors. See figure 8-29. In the northern parts of the U.S. country, pigs must be raised

Photo by Frank Flanders

FIGURE 8-29

Pigs can be raised outdoors if allowed access to well-bedded shelter during cold weather.

with access to shelter during the coldest part of the year. Shelter from inclement weather is very important if pigs are raised outdoors. They can tolerate cold weather to a point, but the amount of feed must be increased to maintain body temperature. The growth rate of pigs slows if the temperature drops below 65 degrees F for growing pigs and 60 degrees F for finishing pigs. Pigs must have a warm, dry, well-bedded place to lie down and keep warm during cold weather especially if no supplemental heat is available. Sows can farrow in outdoor huts in cold weather if heat lamps are provided. Most sows are farrowed in environmentally controlled buildings that are heated to a constant temperature.

Pigs tolerate cold better than excessive heat. Pigs reduce feed intake when the temperature climbs over 80 or 85 degrees F, thus reducing growth rate. This reduction in feed consumption is particularly hard on lactating sows, which often lose excessive body weight and fail to rebreed after weaning. Very high body temperatures render boars infertile. Pigs do not have functioning sweat glands, so a cooling mechanism must be provided. Indoor confinement buildings normally have a sprinkler system installed that drips water onto pigs when the temperature reaches 85 degrees F. See figure 8-30. Sprinklers combined with rapid air movement helps keep pigs comfortable except in the hottest days of summer. Outdoor facilities should have shade and access to a wet place for pigs to lie down and keep cool during the hottest part of the day.

In contemporary swine housing systems, pigs are raised on slotted floors so that manure and urine drop into a reception pit underneath the building. In deep pit buildings, the manure stays there until it is removed for application as fertilizer onto farm fields. In other cases manure flows to an outdoor holding pond for storage. Long-term manure storage facilities are engineered to exacting specifications so that manure does not come in contact with ground or surface water. See figure 8-31.

Photo courtesy of Robert Meinen

FIGURE 8-30

Sprinkler systems in environmentally controlled buildings keep pigs cool and comfortable during hot weather.

Animal Science Facts

In 1950, each American consumed an average of 12.6 pounds of lard (rendered pig fat) each year. Because of health concerns and replacement fat sources such as vegetable oils, current consumption hovers around two pounds per year.

Photo courtesy of Natural Resource Conservation Service

FIGURE 8-31

In commercial swine operations, manure is normally handled as a liquid.

MARKETING

The sale of market hogs is the culmination of the swine production process. To receive top price, producers must provide what the buyers (packers and consumers) want. This means uniform, lean, heavily muscled pigs with high-quality lean. See figure 8-32. Market hog price is typically quoted as dollars per hundred pounds of carcass weight. This price can easily be translated to live price by multiplying carcass price by the dressing percentage (percent of hogs live weight remaining after slaughter). This figure is usually about 75 percent. So if carcass price is quoted at $60/ hundred pounds, the equivalent live price would be $45 per hundred pounds ($60 × 0.75). Prices quoted in $/hundred pounds can easily be converted to cents per pound by dividing the price by 100. For example, a live price of $45/hundred pounds equals $0.45/pound.

Leanness can be traced to genetics and feed. Muscle quality is also a function of genetics. Producers determine size and uniformity of market

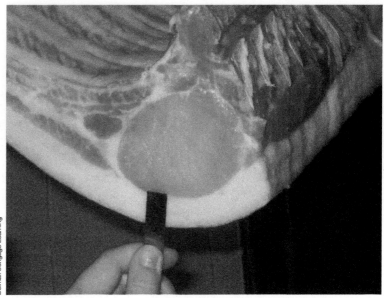

Delmar/Cengage Learning

FIGURE 8-32

Fat depth, muscle quantity, and muscle quality can all be evaluated from the loineye muscle of a hog carcass.

hogs. Packers want pigs to fall within a given weight range (for instance 250 to 280 pounds) so that all the hams, loins, and other cuts of meat are of similar size. Therefore, packers penalize producers if pigs do not fall within the accepted weight range. This penalty is called a sort penalty. Moreover, pigs may receive a price bonus or penalty depending on the fat thickness of the carcass, even when sold at an acceptable weight.

Producers use certain strategies to receive more bonuses and fewer penalties for weight and fat thickness. The first is to weigh all pigs going to slaughter. An accurate scale helps producers find the precise weight of the pigs. Alternately, self-sort finishing barns require all pigs to travel through scales on their way to the feeding area. Pigs heavy enough to be marketed are automatically directed to an adjacent pen. Self-sort technology can be used with different set-point weights if barrows and gilts are housed in separate pens. The second strategy is to sell barrows at the lighter end of the weight range, since barrows get fatter at lighter weights than gilts. For example, if a barrow is sold at 240 pounds, it still may be lean enough to garner a premium.

Photo courtesy of Robert Meinen

FIGURE 8-33

Finishing pigs have access to a food court dining area in a large pen self-sort finishing building.

However, if the same barrow was allowed to reach 270 pounds, it could be penalized for being too fat. Gilts are more likely to stay lean at heavier weights and can be slightly heavier than barrows when sold. See figure 8-33.

CASH PRICE

Cash price on a given day is the live price paid by packers for market hogs. The cash price is determined by supply of, and demand for, pigs on a given day. The cash price is a base price. An individual pig may be worth more or less than the base price depending on bonuses or penalties. For example, the cash price may be quoted at 40 cents per pound. An individual pig weighing 250 pounds would be worth $100 ($0.40 × 250). If the pig is leaner than average, it may receive a two cent per pound bonus. In that case, the pig would be worth $105 ($0.42 × 250). If the pig was fatter than average, it may receive a three cent per pound penalty. This fat version of the 250-pound pig would be worth only $92.50 ($0.37 × 250). The lean 250-pound pig is worth $12.50 more than the fat 250-pound pig.

FUTURES PRICE

Futures price is the expected cash prices at a certain point in the future. There is no guarantee of what the cash market will be on that date, but producers can use futures price as an insurance policy to guarantee a minimum price sometime in the future. If a producer has pigs to sell in three months, he or she can guarantee a price for that time according to the futures market. This means that when three months pass, the producer will receive the price set three months ago, regardless of the cash price on the day of sale. Producers use futures as a way to reduce the risk of unpredictable market fluctuations. Futures and current cash prices can be found on the Internet, in local farm papers, on the radio or television, or through various data services. Futures prices are always quoted on a carcass weight basis ($/hundred pounds of carcass weight).

FORMULA PRICE

Formula price is a prenegotiated price determined between the producer and the packer. Regardless of current market conditions, the producer receives a price based on current cost of production, a price window (designating a minimum or maximum price), or other formula. This formula may designate premiums and penalties depending on the leanness, muscling, or quality of the pigs.

INDUSTRY ORGANIZATION

The swine industry can be classified by operation type. Four basic types of operations exist. **Seedstock** producers or companies provide registered and nonregistered or planned crossbred breeding stock. Producers purchase seedstock hogs to better the genetics, and therefore, the profitability of the herd. Seedstock operations concentrate on improving economic traits that their customers desire, such as growth rate, feed efficiency, carcass traits, or litter size. Farrowing operations maintain sows and sell young pigs to finishing operations. Farrowing operations would be the customers of seedstock producers. Finishing operations feed hogs until slaughter—five to six months of age and weighing 250 to 280 pounds. See figure 8-34. Many producers have **farrow to finish** operations where hogs are owned from birth to slaughter.

Photo courtesy of Robert Meinen

FIGURE 8-34

A contemporary swine finishing facility that houses approximately 2,200 pigs.

INDUSTRY ISSUES

Sustained profitability remains the key issue in the swine industry. Since the 1990s, producers have seen drastic swings in both hog and feed prices. There have been periods of healthy profits interspersed with periods of extreme economic losses. Fully independent producers have virtually disappeared. Some smaller producers who formerly sold breeding stock now remain viable by selling show pigs and pig semen to youth. Nonetheless, many producers were forced out of the swine production business. Larger, integrated firms that could more easily withstand price fluctuations and take advantage of economies of scale stepped in to fill the void in swine production.

Environmental concerns surrounding manure handling face the swine industry. Most swine manure is stored in pits and lagoons where little decomposition can occur. When the manure is applied to cropland, odors can impact neighboring areas and manure nutrients have the potential to reach waterways. See figure 8-35. If manure is incorrectly managed, water pollution can result from manure runoff and excessive leaching through the soil. State and federal regulations require larger swine operations to obtain an approved nutrient or manure management plan that details where, how, and when they will store and apply manure. Producers have to maintain records pertaining to nutrient management. As a result of exceedingly tough environmental standards, permits for new, large-scale operations are becoming increasingly difficult and expensive to obtain.

Product safety and animal welfare concern pork consumers. Most packers require their pork producers to maintain Pork Quality Assurance (PQA) certification from the National Pork Board. This educational program leads producers through 10 Good Production Practices to ensure pork quality. Large eating establishments pressure packers to verify that the pork they sell was raised under acceptable production conditions. The most recent

Photo courtesy of Robert Meinen

FIGURE 8-35

Swine manure is a source of valuable nutrients for field crops.

version of the Pork Quality Assurance program (PQA PLUS) includes a facility assessment and contains enhanced education to ensure animal welfare. See figure 8-36. In some states animal rights activists have spearheaded ballot referendums to prohibit sow gestation stalls.

More concentrated swine production means a greater potential for disease transfer within a production facility. Therefore producers take great pains to ensure that diseases are kept out of facilities by practicing good biosecurity. The threat of foreign animal disease introduction and bioterrorism strengthens the need for tight biosecurity practices at swine farms.

CAREER OPPORTUNITIES

Career opportunities exist for students interested in the swine industry. While some industry jobs require a college education, many can be obtained with a high school diploma. Companies often hire workers and unit managers to conduct the daily operations of swine barns. Tasks for these employees can include checking feeders and waterers, monitoring pig health,

PQAPLUS
Our Responsibility. Our Promise.

pork.org or (c) the Pork Checkoff Service Center @ (800) 456-PORK

pork checkoff

Photo courtesy of the National Pork Board

FIGURE 8-36

The National Pork Board's PQA PLUS program includes animal welfare education for producers.

record-keeping, breeding, farrowing, and performing facility or equipment repairs. Many swine companies also employ supervisors to provide coordination and leadership to specific divisions like finishing or genetics. Large companies may employ a herd veterinarian as well.

Sales personnel market swine feeds manufactured by a multitude of feed companies. Also, interested individuals may choose to work for one of the many swine housing or equipment companies.

Due to the integration of the industry, transportation to slaughterhouses may be provided by the producer or an independent trucking company with dispatchers and drivers. See figure 8-37. Packing firms, too, can be owned by producers. However, most firms operate independently. Thousands of employees are needed to process the millions of hogs slaughtered each year!

Construction and ventilation experts are essential to maintain air quality and manure storage integrity of large-scale swine operations. Nutrient management experts are often employed to dictate where and when manure is applied on nearby cropland.

The swine industry relies heavily on researchers to provide insight into more efficient swine production.

Photo by Robert Meinen

FIGURE 8-37

Hog transportation provides many jobs in the swine industry.

Extension agents, government workers, and teachers disseminate this new information to other swine industry personnel. Furthermore, communication specialists help producers market their product to consumers with advertising slogans, such as, "Pork . . . The Other White Meat."

SUPPORTING ORGANIZATIONS

From a slate of producer nominees, 15 individuals are appointed by the U.S. Secretary of Agriculture to the National Pork Board. This group administers funds gleaned from the pork **check-off** program, which began in 1986. The intent of the check-off program is to finance education, research, and product promotion. Pork check-off monies are collected at the rate of 40 cents per $100 of gross price. Part of the collected check-off monies is returned to the states for promotion, using a variable percentage scale based on number of hogs sold. See figure 8-38.

Housed in Des Moines, Iowa, the National Pork Board provides a voice for producers across the country and develops educational materials, promotions, and advertising for pork products and

FIGURE 8-38

Pork check-off dollars assist the industry in developing promotional campaigns that entice consumers to eat pork.

Photo courtesy of USDA

pork producers. The National Pork Board also hosts the Pork Forum each spring. Producers voice current concerns and set industry policy at this conference. The National Pork Board and the state affiliate organizations are supported by pork check-off dollars.

The National Pork Producers Council (NPPC) hails from Washington, DC, and sponsors the World Pork Expo, which is held annually in June. Participants from around the world attend this industry-oriented event. The NPPC tracks legislative issues for pork producers nationwide.

SUMMARY

The swine industry has changed dramatically over the past few decades. Although eight purebred swine breeds serve as the base for the industry, varying breeding systems use these genetics to assure heterosis (hybrid vigor). Moreover, management practices have changed over time to meet the needs of the industry's vertical integration. Profitability remains a core issue in the swine industry. Looking ahead, environmental issues and consumer demands will help shape future swine production.

CHAPTER REVIEW

EXPERIENTIAL LEARNING OPPORTUNITIES

1. Contact a local Cooperative Extension Educator or visit the National Pork Board Web site to find an individual who delivers the Pork Quality Assurance program. Arrange to participate in the course. Doing so will give a perspective on the industry's efforts to improve food safety and quality.

2. Swine integrators often list positions for laborers. Identify an integrated swine company and search its Web site for employment opportunities.

DEFINE ANY TEN KEY TERMS

all-in all-out production	cross-fostered
atrophic rhinitis	farrow to finish
backcrossing (crisscrossing)	formula price
biosecurity	futures price
cash price	independent producer
check-off	integrated production

leptospirosis

maternal breeds

mycoplasmal pneumonia

parvovirus

paternal breeds

PRRS

pseudorabies

rotaterminal crossbreeding

seedstock

split-sex feeding

swine erysipelas

three-breed rotational cross

terminal crossbreeding

wean to finish

QUESTIONS AND PROBLEMS FOR DISCUSSION

1. Classify the eight major breeds of swine as being maternal, paternal, or combinations breeds.

2. Which three breeds of swine are predominantly black?

3. Which extremely lean breed of pig was introduced into the United States from Germany and Belgium?

4. Only one maternal and one paternal breed are used in the _____ or _____ system.

5. Why is the three-breed rotational cross superior to the backcrossing or crisscrossing system?

6. How many maternal breeds are required for the rotaterminal crossbreeding system?

7. Which breeding system retains no gilts for replacements?

8. What are the six steps involved in processing baby pigs?

9. Differentiate between nursery, grower, developer, finisher, gestation, and lactation diets.

10. List two external swine parasites.

11. Why is it essential to provide pigs with a cooling mechanism if the temperature rises above 85 degrees F?

12. Why do packers penalize producers with a sort penalty for pigs that do not fall within a specified weight range?

13. Between barrows and gilts, which remain leaner at heavier weights?

14. Packers often give bonuses for _____.

15. What is the current cash price for a market hog?

16. Which region of the United States is the heart of swine finishing operations?

17. In the early 1960s over _____ American farms produced an average of 50 hogs per year.

18. Smaller farms of less than 1,000 head house _____ percent of the swine inventory in the United States.

19. Feed accounts for about _____ percent of total swine production costs.

20. Name the type of hog operation that sells breeding stock?

21. Hogs are owned from birth to slaughter on a _____ to _____ operation.

22. True or False. Large, integrated firms can more easily withstand extreme hog and feed price volatility than small independent producers.

23. List two environmental concerns surrounding manure handling.

24. _____ hog producers operate without company affiliation.

25. What does PQA stand for?

26. The pork check-off program began in _____.

27. Pork check-off dollars are collected at a rate of _____ of gross price.

28. In which city is the National Pork Board located?

29. Which large-scale swine industry event is held annually in June?

30. Write the famous pork advertising slogan.

CHAPTER

9

Beef Cattle Management and the Beef Industry

HOME ON THE RANGE

Objectives

After completing this chapter, students should be able to:

▸ Give an overview of the beef industry in the United States

▸ Describe the physical characteristics of the major beef breeds

▸ Explain the relationship between the seasons and beef production

▸ Discuss nutrient requirements for various stages of beef cattle production

▸ List and describe parasites and diseases common to breeding and market cattle

▸ Discuss the housing requirements for beef cattle

▸ List three factors that affect the cost of feeder calves

▸ List four production segments of the beef industry

▸ Discuss two issues facing the beef industry

▸ List three careers in the beef industry

▸ List two organizations that support the beef industry

Key Terms

backgrounding	eared cattle	performance testing
baldies	feeder calf	polled
brood cow	finisher	progeny
commercial cattle	finishing cattle	replacement heifers
compensatory gain	grower	settle
continental breed	intensive rotational grazing	shipping fever
cow-calf		terminal sire
dressing percentage	lactating	weaning
dystocia	marbled	winter

Career Focus

Cody recalls coming home from school one day to find his FFA steer not feeling very well. The calf's nose was running, it was standing in the back of the pen with its head down, and it was not interested in eating. Cody called the veterinarian Dr. Britt who arrived shortly afterward. Dr. Britt quickly discovered that the steer was running a fever and prescribed some antibiotics to treat the steer for pneumonia. After the examination, Dr. Britt and Cody tried to find where the steer could have acquired this disease. Cody relayed that he had taken the steer to a jackpot show the prior weekend. Dr. Britt then asked Cody if the calf had been vaccinated for shipping fever. Cody did not know for sure, but was certain the steer had not been vaccinated since he bought it five months ago. Dr. Britt suggested that vaccinations are important to prevent respiratory disease in finishing cattle.

This incident stayed with Cody through the remainder of his high school career, through a bachelor's degree in Animal Science. In fact he told this story during a job interview that landed him a new position as an animal pharmaceuticals representative servicing large feedlots in the panhandle of Texas.

INTRODUCTION

Delmar/Cengage Learning

FIGURE 9-1

Beef cattle can convert forages into high-quality meat.

Beef cattle are different from hogs because cattle can convert forages, nutritionally of little use to humans, into valuable, high-quality protein. Much of the land pastured for beef cattle is too steep, rocky, or wet to be cultivated by conventional methods. Beef cattle make excellent use of this land. Beef cattle can be successfully wintered outdoors while consuming poorer-quality forages than those required by dairy cattle. Extremely hardy creatures, beef cows can survive and produce a calf with minimal management, making them ideal for part-time producers. Beef cattle are indeed "at home on the range."

Beef animals are raised throughout the United States because of their ability to convert forages to meat and the low overhead costs needed for their production. Cattle production ranks first in U.S. livestock sales receipts, accounting for approximately 21 percent

of total farm income and 40 percent of all animal sales receipts. Although larger beef cattle operations can be segmented into four basic production types, many farmers raise beef, from calving until slaughter, for personal consumption and extra income. However, the beef industry does not stop at the slaughterhouse door. Thousands of employees are needed in the packing, inspection, distribution, promotion, and retailing of beef.

INDUSTRY OVERVIEW

Cattle generate more dollars than any other agricultural enterprise. Twenty-one percent of total farm income comes from the beef cattle industry. Texas ranks first in beef cattle production with Missouri, Oklahoma, Nebraska, and South Dakota following in places second through fifth, respectively. These five states account for nearly 40 percent of total beef cattle in the United States (2007 USDA data).

Over 765,000 cow/calf producers in the United States average approximately 552 acres of owned and rented land. An additional 50,000 producers reported having cattle on feed in the 2007 USDA census of agriculture. In the United States, land utilized for pasture and range totals nearly 445 million acres. Most beef cows and calves are located in the southeastern and western states, while states on the high plains form the center of feedlot operations. More small, part-time beef producers operate in the east, while western producers tend to run large full-time operations.

When market prices are high, most animal enterprises experience an expansion in numbers. More inventory typically causes prices to fall and national numbers to contract. Typically, beef cattle numbers cycle every 10 years because of the time required to hold back replacement heifers, breed them, and work their progeny through market channels. However, a variety of reasons—including long-term drought in many cow/calf production areas—have slowed the most recent rebuilding cycle.

BREEDS

Beef breeds come in an amazing variety of shapes, sizes, and colors. Most beef breeds originated in Europe from the native stock first domesticated there.

LONGHORNS

The first American breed of beef cattle was the Texas Longhorn, introduced by the Spanish to the Southwest in the 1500s. Until the mid-1800s, the Longhorn was the only beef breed available to the majority of American cattle producers. Longhorns were hardy creatures—a necessity at that time—but lacked a beefy appearance and were difficult to fatten. See figure 9-2. The importation of Shorthorn cattle, and subsequently Herefords and Angus (all from the British Isles), to cross with the native Longhorns gave **commercial cattle** (crossbreds) heavier-muscled carcasses. Other advantages of these three British breeds, including increased growth rate, improved mothering ability, and shorter horns, drove Longhorns to the brink of extinction. However, a recent increase in the popularity of Longhorns due to their lean carcasses and for use as roping stock has boosted 2008 registrations to 10,500.

Photo courtesy of the Texas Longhorn Breeders Association / Cole Dowden

FIGURE 9-2

Longhorns were the first cattle of the American West.

For the better part of 100 years, Shorthorns, Herefords, and Angus ruled the American cattle industry.

SHORTHORNS

Shorthorns were the first breed imported from Britain to the United States in the late 1700s. See figure 9-3. The Shorthorn name derives from the breed's ability to shorten the Longhorn's lengthy horns. Shorthorns can be red, white, roan, or red and white spotted. Because they are early-maturing and grow more rapidly than the Longhorns, Shorthorn cattle can be given credit for stamping a beefy appearance on American cattle. Shorthorns have seen a resurgence in popularity. Further the breed supports a healthy youth organization. Shorthorn registrations totaled 15,715 in 2008.

HEREFORDS

During the early 1800s, Herefords became the second British breed to be imported to the United States. See figure 9-4. Herefords are red with white faces, bellies, and tail switches. White markings on the back of the neck are also common. Hereford crossbreds retain the characteristic white face and are often called

FIGURE 9-3

This Shorthorn cow probably looks much different than the original imports to the United States.

Photo courtesy of USDA

FIGURE 9-4

Herefords have left their trademark white face on many of America's crossbred cattle.

baldies. Hereford and Angus crosses are known as black baldies, exhibiting black bodies and white faces. Herefords are extremely hardy animals that are able to tolerate a wide range of environmental conditions. In 2008, the American Hereford Association registered 63,943 cattle.

In the early 1900s, Warren Gammon located ten registered Hereford cows and one registered Hereford bull that were naturally **polled** or hornless. See figure 9-5. These 11 cattle were the foundation of today's Polled Hereford breed. Polled Herefords are used much the same as Herefords with the American cattle producers, but the polled gene eliminates the task of dehorning calves.

BLACK ANGUS

Aberdeen Angus, or more simply Angus, were first imported from Scotland in the late 1800s. See figure 9-6. Angus are by far the most popular breed of beef cattle with registrations totaling 333,766 in 2008. Angus are usually black in color and always naturally polled. Angus breeders have paved the way for industry-wide **performance testing**, which generates growth rate data and information on other economically important genetically inherited traits.

Photo courtesy of the American Polled Hereford Association

FIGURE 9-5

Polled Herefords were developed from a strain of naturally hornless Hereford cattle.

Photo courtesy of the American Angus Association

FIGURE 9-6

Maternal strength, structural correctness, and documented performance combine to make Black Angus the most popular American beef breed.

RED ANGUS

Some Angus possess a recessive gene for red coloration. See figure 9-7. When two Angus carrying a single copy of the red gene are mated, there is a 25 percent chance of producing a red calf. If two red Angus are mated, all the calves will be red. In the 1950s, a separate breed association was formed for the registration of Red Angus cattle. Since then,

Photo courtesy of the Red Angus Association of America

FIGURE 9-7

Red Angus were developed by isolating the recessive red gene found in Black Angus.

Red Angus have been registered as a distinct breed. In 2008, registrations of Red Angus totaled 48,061.

Until the mid-1900s, these three breeds comprised most of the American beef cattle industry. However, two new breeds arrived during the mid-1800s and early 1900s that were not given much attention at the time. Zebu or Brahman cattle were imported in the mid-1800s, and the Charolais in the 1930s. Neither was seriously considered as a threat to the three established breeds.

BRAHMAN

Brahman cattle are actually a different species of cattle (*Bos indicus* rather than *Bos taurus*) than the other beef breeds. See figure 9-8. They were first exported to the United States from India—their native home. This exportation occurred in several waves from the mid-1800s to the early 1900s. Purebred Brahmans

Photo courtesy of American Brahman Breeders Association

FIGURE 9-8

Loose skin increases the surface area of Brahman cattle allowing heat to escape. This feature makes Brahmans well suited for hot climates.

have loose hides and long ears, which continue to be evident in their crossbred offspring. For this reason, crossbred or purebred Brahman cattle are sometimes called **eared cattle**. A hump over the shoulders and a relatively long face, as compared to other breeds, are also trademarks of Brahmans. Purebred Brahmans are usually gray, but other colors can also be found. Brahmans are usually found in hot climates because their loose hides dissipate heat efficiently. Brahmans are also more tolerant to diseases than many other breeds of cattle. They have been used as a foundation breed for the development of several new breeds of cattle. Registrations of purebred Brahmans numbered 8,500 in 2008.

Several distinctly American breeds of cattle rely on a high percentage of Brahman blood. Brangus (5/8 Angus, 3/8 Brahman), Beefmaster (1/2 Brahman, 1/4 Hereford, 1/4 Shorthorn), and Santa Gertrudis (5/8 Shorthorn, 3/8 Brahman) are very popular breeds in the South and Southwestern ranges where their hardiness and maternal abilities are well known. See figure 9-9. Registrations for these breeds numbered 29,643, 14,692, and 7,500, respectively, in 2008.

Photo courtesy of International Brangus Breeders Association

FIGURE 9-9

Brangus cattle combine the performance of Angus with the heat tolerance of Brahman cattle.

Photo courtesy of the American Charolais Association

FIGURE 9-10

Heavily muscled Charolais bulls such as this one are used to sire thick, fast-growing calves.

CHAROLAIS

Charolais (shar-lay) cattle originated in France as draft animals. See figure 9-10. Fully white in color, Charolais are noted for their rapid growth rate and muscling. The first imports came from Mexico in the 1930s, but the breed association was not established in the United States until the early 1950s. The Charolais was the first imported **continental breed** (so-called because it came from the continent of Europe, not the British Isles). Cattle breeders were amazed with the growth rate and muscling of Charolais crossbreds. However, the early imports tended to sire large calves that were born with difficulty. Many other continental breeds displayed the same trait. American breeders have judiciously selected breeding stock for low birth weights, therefore, genetically solving most of the birth weight problems. Total active registrations of Charolais cattle numbered 65,954 in 2008.

In the 1970s, imports of other continental breeds mushroomed. The availability of these new breeds was often limited to small numbers of bulls available only through artificial insemination. To rapidly increase registration numbers, breed associations allowed producers to breed cattle up to purebred status. This involved breeding purebred bulls to crossbred (or purebreds of another breed) cows and registering the **progeny** or offspring as half-bloods. The half-blood progeny were then bred to another purebred bull, and the resulting 3/4 bloods were also registered. Many continental breed registries consider 7/8 blood or 15/16 blood animals to be purebreds. Depending on the breed of the original cows in the breeding up process, black and/or polled status could be maintained in a purebred animal from a continental breed that was originally red and horned. Three examples, Limousin, Maine-Anjou, and Simmental, will be discussed later in the chapter.

American cattle producers recognized the advantages of crossbreeding and heterosis. Newly imported breeds arrived in large numbers and breeders quickly incorporated them into crossbreeding programs. Presently, cattle producers have a vast array of genetics from which to choose including the original British breeds, as well as many newer breeds. A brief description of the more popular continental breeds used in the United States follows. This list is not comprehensive.

CHIANINA

Chianina (key-a-nee-na) are the largest breed of beef cattle with mature bulls standing six feet high and weighing up to two tons. See figure 9-11.Originating in Italy as a draft animal, purebred Chianinas are white

Delmar/Cengage Learning. Courtesy of the American Chianina Association

FIGURE 9-11

Large-framed Chianinas are known for their lean carcasses and rate of gain.

to gray in color with black, pigmented noses, tails, and hooves. The white color is recessive to any other color in a Chianina crossbred animal. In other words, a white Chianina mated to a black or red animal will always be black or red. Chianinas were first introduced to the United States through semen imports in the early 1970s. They are noted for their growth rate and lean carcasses. Registrations of Chianina cattle totaled 9,756 in 2008.

GELBVIEH

Gelbvieh (gelb-fee) cattle originated in southern Germany and were first introduced to the United States in the early 1970s. See figure 9-12. Gelbviehs are solid red or black. The American Gelbvieh Association requires individual performance data for birth weight and growth before animals can be registered. In 2008, 37,448 animals were registered.

LIMOUSIN

Limousins are a French continental breed first introduced to the United States through semen imports in the late 1960s and early 1970s. See figure 9-13. Originally red in color and horned, black and polled lines have been developed by American breeders. Limousins are moderately framed cattle known for their heavily muscled carcasses and rapid growth rates.

Photo courtesy of the American Gelbvieh Association

FIGURE 9-12

Functional, performance-oriented Gelbvieh cattle are extremely useful to commercial producers

Photo courtesy of the North American Limousin Foundation

FIGURE 9-13

Limousin cattle originated in France and are known for their exceptional muscle development and carcass quality.

They have proven to be an excellent choice for crossbreeding with large-framed or light-muscled cows. There were 28,928 Limousins registered in 2008.

MAINE-ANJOU

Maine-Anjou (main an-ju) cattle were imported from France in the early 1970s. See figure 9-14. Known for their maternal abilities and docile temperament, Maines

Photo courtesy of the American Maine-Anjou Association

FIGURE 9-14

Maine-Anjou cattle are another French breed known for their muscularity, growth rate, and docility.

Courtesy of the American Salers Association

FIGURE 9-15

Salers cattle have a reputation for exceptional performance.

are the largest framed and heaviest of the French breeds. They were originally red and white, but as with Limousins, American breeders have developed black and polled lines. Maines are also popular in terminal crossbreeding programs and as show steer sires. Registrations of Maine-Anjou cattle reached 10,368 in 2008.

SALERS

Salers (sa-lair) are a third French breed imported shortly after Maines. See figure 9-15. Salers are dark red and may have white tail switches and belly spots. Salers have a reputation among commercial cattle producers as a performance-oriented breed. In 2007, 13,667 Salers were registered.

SIMMENTAL

Simmental cattle were developed as a true multipurpose animal in their native Switzerland. See figure 9-16. Large and heavily muscled, they could work as draft animals, produce large amounts of milk, and have acceptable carcasses when slaughtered. The first Simmentals came to America in 1971, but semen was available several years before. Simmentals were originally red or yellow and white with a prominent white face that passes to crossbred progeny. Black and polled lines are available. Simmentals can be successfully used as a maternal breed or as a **terminal sire** (father in a crossbreeding system where all progeny are marketed). The versatility of the Simmental explains its popularity with 45,500 registrations posted in 2008.

(Numbers of registered animals for individual breeds obtained from National Pedigreed Livestock Council http://www.nplc.net/beef.html.)

Photo courtesy of the American Simmental Association

FIGURE 9-16

Strong maternal characteristics typify Simmental cattle.

© Dale A. Stork. Used under license from Shutterstock.com

FIGURE 9-17

Consistent group of replacement heifers for a commercial beef operation.

BREEDING SYSTEMS

Breeding systems are not as well defined in beef cattle as in swine. Many commercial producers have essentially purebred herds. Others may use a loose rotational crossbreeding program. In the future, beef producers must identify specific crossbred cows that wean heavy calves, rebreed easily, and produce for many years. These cows should be bred to heavily muscled, fast-growing bulls from early-maturing bloodlines. Many producers may purchase **replacement heifers** (young breeding females) of specific crosses rather than raise their own. See figure 9-17.

BEEF PRODUCTION CYCLE

The beef production cycle is a year-long affair that begins at calving time. In most parts of the United States, calving starts just before the grass begins to turn green in spring (as early as January in southern states to April or May in northern states). However, some producers keep fall calving herds to take advantage of the higher-priced **feeder calf** (young cattle just weaned) market brought on by a shortage of feeder calves in the spring.

Management at calving is crucial to reproductive success. A live calf is the only product beef producers can sell. Therefore, producers make a special effort to help ensure all calves are born alive and healthy. Calving should occur in clean pastures with enough room for the cow to get away from the herd. Producers check cows frequently during the calving season to assist cows that may have calving problems (**dystocia**). Heifers calving for the first time, small cows with big calves, and abnormal positions of the calf are the most common causes of dystocia. Experienced managers know when to assist the cow. Calves should be standing and nursing within one or two hours of birth. Disinfecting the naval with iodine, tagging or tattooing for identification, and castrating the calf are management practices performed within 24 hours of calving. See figure 9-18.

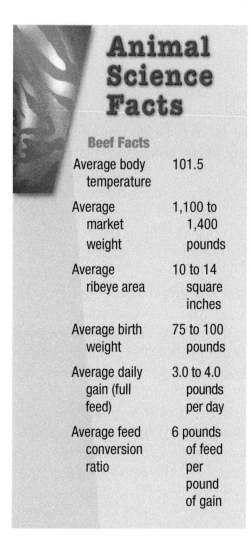

Animal Science Facts

Beef Facts

Average body temperature	101.5
Average market weight	1,100 to 1,400 pounds
Average ribeye area	10 to 14 square inches
Average birth weight	75 to 100 pounds
Average daily gain (full feed)	3.0 to 4.0 pounds per day
Average feed conversion ratio	6 pounds of feed per pound of gain

Lactating cows (those giving milk) need large amounts of high-quality forage. Spring growth of most pasture and range grasses contains high levels of energy and protein, which allow cows to give milk, build body condition, and subsequently be rebred for the next year. See figure 9-19. Since nutritional demands during lactation are low when the calf is first born, and peak about two to three months into lactation, calving is often timed about a month before cows can be turned

FIGURE 9-18

Calves should be up and nursing within an hour or two of birth.

FIGURE 9-19

Lactating cows need high-quality forages to produce milk and rebreed.

onto grass. Thus, the peak of lactation occurs when the highest-quality forage nutrients are available.

The breeding season should begin 60 to 75 days after calving and should last about 60 days. This season normally occurs in early to midsummer depending on latitude and elevation. Heifers being bred for the first time should be bred about 30 days before mature cows for two reasons. Heifers will deliver the first calves of the season. Therefore, they can receive more attention at calving time. Also, this schedule allows first-calf heifers an extra month to rebreed. This is necessary since first-calf heifers are often the most difficult cows to **settle** (become pregnant) because of the added nutritional demand of their own growth and the demands of lactation. Gestation length for cows is approximately 280 days. Therefore, next year's calf should be born at roughly the same time as this year's calf [gestation (280 days) + pre-breeding period (60 days) + average time before breeding (30 days) = 370 days].

Producers usually restrict the calving season by introducing and removing bulls in a timely manner. A 60-day calving season helps keep groups of calves uniform in size, an economically important factor at marketing time. A short calving season also concentrates the period of intense labor. A 60-day calving season can be accomplished by introducing the bull(s) and removing him (them) 60 days later. Pregnancy checks can be performed 50 to 60 days after the removal of the bull (late summer to early autumn). Any open (not pregnant) cows should be sold.

Weaning (removal from the dam) normally takes place in autumn when grass production ceases. At this time, calves are about six months of age. After weaning, pregnant cows are "wintered" on a diet of lower-energy forage, such as poorer quality hay, until two or three months prior to the next calving season. See figure 9-20. During the last 60 to 90 days before calving, dietary energy levels should be increased to account for the heavy nutritional demand imposed by the rapidly growing fetus.

Bull calves not castrated at birth are usually castrated at weaning and all horned calves are dehorned. Calves not retained for breeding stock are often implanted with a growth promotant. See figure 9-21.

FIGURE 9-20

Dry pregnant beef cows can be wintered on a diet of poorer-quality hay.

FIGURE 9-21

Calves destined for the feedlot can be implanted with a growth promotant.

Weaned calves take one of several possible routes. Some heifer calves may be retained for breeding the following summer. Steer or heifer calves may be placed in a feedlot where they are fed high-energy diets in preparation for slaughter. Calves may also be placed on a low-energy, high-forage, low-cost diet for their first winter, then placed in a feedlot the following spring. This practice is called **backgrounding**.

BEEF CATTLE NUTRITION

Beef cows depend on the nutrients contained within forages for nutrition. As discussed previously, the beef cow needs the most nutritious diet (good forages) when her demands are the highest (during late gestation and lactation). Dry, gestating mature cows require relatively little energy and protein.

Forages come in two types: pasture or range grasses and stored forages, such as hay or silages. During the growing season, pasture or range grasses are the obvious choice for feeding. Pasture and range management varies tremendously in different parts of the United States. For instance, in the East, some pastures are intensively managed to produce as much forage as possible. In an **intensive rotational grazing** system, cattle are allowed access to a paddock of grass for as little as a day or two. Under such a system, 1 to 1.5 acres can support a cow/calf pair for an entire year. See figure 9-22. In drier western states, however, the

FIGURE 9-22

Cattle are moved into small paddocks for short periods in an intensive rotational grazing system.

lack of moisture and alkaline soils will not support such intensive management. Beef cattle are raised in areas that require up to 110 or more acres per cow/calf pair.

Stored forages are normally fed to dry, pregnant cows in winter when range and pasture grasses are dormant. Free-choice hay is perhaps the most commonly stored forage. Depending on body size, dry, pregnant cows consume approximately 25 to 30 pounds of hay per day. Pregnant heifers should be fed better quality hay than mature cows since they have not yet reached mature body weight.

Bulls can be fed stored forages when not breeding. They may require some grain before, during, and after the breeding season to maintain body condition.

During their first winter, heifer calves are often fed a grain ration in addition to stored forage. Five to seven pounds of a grain mix, similar to that shown in figure 9-23, coupled with good-quality, free-choice hay will ensure heifers are heavy enough (at least 60 percent of expected mature body weight) for breeding at 13 to 14 months of age. Bulls, cows, and

Feedstuffs	Heifers	Low Energy	High Energy
Ground ear corn	95		
Shelled corn		55	99
Oats		35	
Soybean meal		7	
Salt, vitamins, minerals	5	3	1
Total	100	100	100
Amount to feed/hd/day	3 to 6 lb	6 to 18 lb	18 to 25 lb
Amount of good-quality hay	Free choice	6 lb	4 lb

1. Total feed consumption should be about 2.5% of live weight.
2. Salt, vitamin, and mineral recommendations are approximate. Consult your instructor or follow directions of a commercial premix.
3. The amount of grain should be increased gradually, starting at 2 to 3 lb per day and increasing by 1 lb every 2 days.
4. Heifer diet is based on 500- to 800-lb calves. Low-energy diet is for 500- to 900-lb calves. High-energy diet is for finishing cattle 900 lb and above. High-energy diet can be fed to large-framed lighter calves for the entire feeding period at lesser amounts and with higher levels of vitamins and minerals.

Delmar/Cengage Learning

FIGURE 9-23

Rations for heifer growing, and low-and high-energy diets for finishing cattle.

FIGURE 9-24

Feedlot cattle eat a high-energy diet to fatten for slaughter.

bred heifers should have access to free-choice salt and minerals at all times.

Finishing cattle (feedlot cattle) are normally fed a high-grain, high-energy diet containing very little forage. See figure 9-24. This results in rapid gains and increased carcass quality. The amount of grain calves receive when they first enter a feedlot is relatively low—only three pounds per day, with the rest of the diet made up of forage. The amount is gradually increased until 80 to 90 percent of the diet is grain mix or high-energy corn silage. Young feedlot cattle require some supplemental protein, but as they near slaughter weight, the protein content in the diet can be reduced. Monensin and Lasalocid (ionophores) are two feed additives that increase weight gain and improve feed efficiency in finishing cattle as well as developing heifers. Frame size affects how finishing cattle should be fed. Because high-energy grains fatten cattle quickly, small-framed cattle should be fed a low-energy grower grain mix for a period of time. Average and small-framed cattle are sufficiently fattened when fed high-energy grain for 90 to 120 days at the end of the finishing period. Large-framed cattle require high-energy grain mixes for the entire finishing period.

Growing heifers and finishing cattle should have salt, minerals, and ionophore (Monensin or Lasalocid) added to the grain mix or available free-choice.

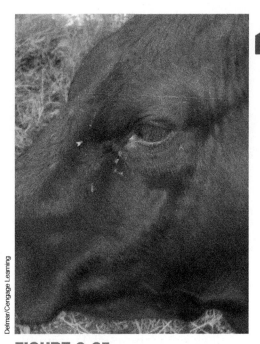

FIGURE 9-25

Flies irritate cattle and can reduce performance.

BEEF CATTLE PARASITES

As with all livestock, cattle are subject to a wide variety of parasites and diseases. A discussion of the symptoms and prevention of the more common or severe parasites and diseases follows.

FLIES

Flies are the most common parasites of beef cattle. Face flies irritate cattle by circling the eyes and feeding from the discharge. See figure 9-25. In addition, horn flies feed on blood and cause irritation by congregating on the belly, back, and sides of cattle. Irritation by either type of flies can reduce the time cattle spend feeding, and thus compromise gains or body condition. Both face and horn flies lay their eggs in fresh manure. They both can be controlled in part by face oiling devices, dust bags containing fly repellent, pour-on products, or ear tags impregnated with insecticide. Reduction of fresh manure piles eliminates fly breeding sites. Heel flies lay their eggs in the hair of cattle. After hatching, the larvae burrow through the hide and migrate to the back of the cow. The larvae, called cattle grubs, mature under the hide—usually during the winter. In spring, the larvae hatch and drop to the ground where they reside until molting into mature flies. The holes left by grubs cause a drastic reduction in the price of the hide after slaughter. Many compounds are available for controlling grubs, including broad-spectrum insecticides, such as ivermectin. However, grub control should be performed only during certain stages in the grub's life cycle.

ANAPLASMOSIS

Anaplasmosis is caused by a parasite that infects the red blood cells of cattle. Most often seen in southern states, anaplasmosis causes lethargy and reduced body condition. Some animals show few signs of parasites but act as reservoirs. Other animals may die. Cattle can be vaccinated against anaplasmosis, but identification and segregation of infected animals are better treatments. Infected animals can be treated with continual antibiotics to prevent the spread of anaplasmosis.

COCCIDIOSIS

Coccidiosis is caused by tiny parasites that infest the lining of the intestine causing severe diarrhea. Cleanliness of surroundings, especially feeding areas, greatly reduces the spread of *coccidia*. Several effective compounds are available for treatment and prevention.

RINGWORM

Ringworm, a fungal infection of the skin, causes round, raised, bald areas on the skin. See figure 9-26. This infection is most common in winter when cattle are often confined in a dark environment. Treatment normally consists of the application of one of many effective solutions. Consult your veterinarian for specific formulas.

MANGE

Mange mites infest the hides of cattle in much the same way as they infest hogs. Treatment consists of the use of one of several pour-on products or complete

Photo by Dr. James Lawhead

FIGURE 9-26

Round, crusty, hairless patches of skin signify ringworm.

Animal Science Facts

immersion of the animal in a treatment solution. Ivermectin is effective in the treatment of cattle mange.

WORMS

Stomach worms routinely infest beef cattle. Oral wormers, as well as broad-spectrum, pour-on products, are very effective on most species. Worm eggs reside at the bottom of grass plants. Overgrazing increases the chances of infestation. Pasture rotation will interrupt the life cycle of the worm.

BEEF CATTLE DISEASES

Diseases affecting beef cattle are divided into three categories: respiratory diseases, reproductive diseases, and gastrointestinal diseases.

RESPIRATORY DISEASES

Several respiratory diseases are grouped under a large umbrella called **shipping fever**, which consists of a complex of three separate diseases, hence the more accurate name is bovine respiratory disease complex (BRDC).

The first segment of the complex is infectious bovine rhinotracheitis (IBR), which is caused by a herpes virus and results in reduced appetite, mucous discharge from the nose, and difficulty in breathing. Treatment is difficult, but vaccination against IBR is fairly standard in the cattle industry.

The second segment of BRDC is bovine viral diarrhea (BVD). Symptoms are similar to IBR with the addition of high temperature and severe diarrhea. Vaccination against BVD is a common practice.

The third segment of the respiratory complex is parainfluenza 3 (PI3). As the name suggests, parainfluenza is a flu-like viral disease often associated with IBR and BVD. Symptoms are interchangeable with those of IBR and BVD.

Vaccination for IBR, BVD, and PI3 can be accomplished with a single vaccine. Initial doses should be given two to four weeks apart, with annual boosters.

REPRODUCTIVE DISEASES

The first reproductive disease of national importance is brucellosis, which is characterized by late-term abortions. Many states are nearly brucellosis free through vaccination combined with comprehensive, statewide testing and removal of infected animals. Brucellosis is a targeted disease for total elimination because it is infectious to humans. The human form of brucellosis is called undulent fever.

Leptospirosis is another easily prevented disease that can cause reproductive difficulties. Leptospirosis can cause abortions at any time during pregnancy and should be included in the vaccination program.

Trichomoniasis is technically a parasitic disease that causes early-term abortions and temporary sterility. Trichomoniasis is spread by infected bulls. This disease can be most efficiently controlled through the use of artificial insemination and uninfected bulls.

GASTROINTESTINAL DISEASES

Often, the first gastrointestinal disease in newborn calves is calfhood scours. See figure 9-27. Like BRDC, calfhood scours can be caused by any of several infectious agents. Bacterial causes include *E. coli*, *Salmonella*, and *Clostridium*, while viral agents include rotavirus and coronavirus. Scouring calves can dehydrate and may die. Treatment with an electrolyte solution to replace lost fluids is generally recommended. Prevention starts with providing a clean place for cows to calve. Adequate colostrum within the first few hours of life is also critical. Vaccines are available but should be used on an as-needed basis depending on the herd history of scours and veterinarian recommendations.

Johne's disease is one of the most difficult and incurable beef diseases to control. A chronic bacterial infection slowly thickens the intestine, preventing proper absorption of nutrients. Infected animals slowly lose body condition while exhibiting a watery diarrhea. Transmission occurs from the feces of infected animals. Although infection

Delmar/Cengage Learning

FIGURE 9-27

Manure caked on the tail and rear quarter identifies a scouring calf.

occurs shortly after birth, symptoms may not be present for several years. Prevention involves removing calves from their dams immediately after birth.

OTHER DISEASES

Tuberculosis (TB) is a wasting disease with symptoms similar to Johne's. Infection may also be present with no outward signs. Animals are infected through contaminated water. Herds should be periodically tested for TB and animals testing positive should be removed.

Other diseases of economic importance include pinkeye, an eye infection easily treated with antibiotics. Foot rot, a hoof ailment caused by wet, infected pastures, is also common. Treatment includes foot baths, antibiotics, and topical treatments.

BEEF CATTLE HOUSING

Housing requirements for beef cattle need not be elaborate. Mature cows need only a sheltered place or windbreak in the most bitter weather. See figure 9-28.

Delmar/Cengage Learning

FIGURE 9-28

A simple windbreak can protect mature beef cattle in the most bitter weather.

Many cattle are wintered in areas without a building. Thick trees or brush can provide adequate shelter. An ideal wintering location should be well drained to keep mud to a minimum in areas where mud is a wintertime problem. In addition, newborn calves should have access to shelter if born during the winter season.

Finishing cattle are fed in various conditions. Small groups of cattle can be fed indoors on a manure pack. Many larger feedlots are entirely outdoors with only a windbreak for shelter and a high dry place on which cattle can lie.

MARKETING

Most cattle in the United States are marketed as calves or as finished cattle for slaughter. Furthermore, some cow/calf producers retain ownership of their calves and pay a feedlot operator to finish them.

Calves can be sold privately at the farm, or through feeder calf sales, tele-auctions, or Internet auctions. Uniformity of size, sex, body condition, and health are important. A history of the carcass value of similarly bred calves can also influence price. Feedlot operators want calves of similar frame size so they can be fed the same way and be expected to finish at the same time. Body condition is important because buyers want calves that are slightly thin. Thin calves that have not been fed much grain exhibit what is called **compensatory gain**. Compensatory gains are very efficient, profitable weight gains of slightly thin calves made at the start of the finishing period. Health is important for obvious reasons. Sick calves are not efficient. Because genetics plays a large role in carcass value, buyers often pay more for calves with proven genetics for carcass composition.

Timing the marketing of finished cattle can make or break a finishing enterprise. Packers want well-**marbled** cattle (intramuscular fat), but with little external fat. Genetics and feeding programs play the biggest roles in desirable fat condition. If cattle have too little fat, the price paid by buyers and packers is greatly reduced. If cattle are too fat, packers have to trim excessive low-value carcass fat. Extra fat also increases the producer's feed costs, since feed efficiency drops dramatically after cattle reach an ideal slaughter point.

Weight of finished cattle is also important. Live weights of finished cattle should fall between 1,100 and 1,350 pounds. Frame size (genetics) and feeding control finished weights. Heifers are sometimes discounted when sold on a live basis because the possibility of pregnancy could severely reduce the **dressing percentage** or percent of live weight.

Cattle from small feedlots are often sold through local auctions or directly to consumers through local custom slaughter plants. See figure 9-29. Larger western feedlots are visited directly by buyers. Bids are taken on a group of cattle and the cattle are sold to the highest bidder. Most larger feedlots sell cattle on a merit system, with high-yielding, highly marbled cattle earning the most money. Alternately many large feedlots contract cattle to specific packers well before slaughter.

As with hogs, cattle can be bought or sold on the cash market or through the futures market. Many cattle are sold on a carcass basis that gives premiums for lean, highly marbled carcasses, and penalties for fat, poorly marbled carcasses.

Photo courtesy of Cathy Collett Livestock Market Assoc.

FIGURE 9-29

Cattle can be marketed through many channels including local livestock auctions.

INDUSTRY ORGANIZATION

Although some producers raise beef cattle from birth to slaughter, most producers can be loosely organized into four types of operations: seedstock, cow-calf, grower, and finisher. Seedstock operations provide breeding animals for cow-calf producers. Seedstock producers develop purebreds (registered or nonregistered animals of a specific breed) or planned crosses to improve such genetic traits as carcass quality or feed efficiency.

Cow-calf operations maintain **brood cows** (mother cows) for an annual calf crop. Weaning of these calves occurs at 400 to 600 pounds at six to eight months of age. At weaning, animals are normally processed (vaccinated, castrated, implanted, dehorned, etc.). Calves then move from the cow-calf producer to either a grower operation, where they mature until approximately 900 pounds, or directly to a feedlot. The main objective of **grower** operations, sometimes called stocker operations, is to grow the animals until they are more mature for finishing. Growers either **winter** the animals in confinement to ready them for placement on spring pasture or background them on a diet intended for more rapid gain and subsequent feedlot placement without additional pasturing. In either case, profits are obtained by adding weight at a lower feed cost than value of the increased weight gain. Grower/stocker cattle go from the grower to the **finisher** (feedlot) for slaughter after a period of high energy feeding. Beef cattle are generally marketed at 14 to 24 months of age weighing 1,100 to 1,300 pounds. The size of feedlots can vary from fewer than 100 to more than 100,000 head.

INDUSTRY ISSUES

Since the death of several fast-food customers, due to *E. coli* poisoning from undercooked hamburgers, and mad cow disease headlines, food safety has been on the forefront of beef industry issues. This issue has been addressed from all angles in the beef production and processing industries. Producers are encouraged to attend Beef Quality Assurance training, and packers and food processors are under increased surveillance and testing programs. While irradiation of ground beef products has been proven to be an effective means of killing bacteria (including *E. coli*) that have accidentally contaminated ground beef products, consumers' fears of the word irradiation have essentially stalled this technology. In addition, industry-wide educational efforts have encouraged consumers to cook hamburger thoroughly (until the center turns gray and juices run clear), and properly clean utensils and surfaces that have contact with raw or undercooked meat.

The cattle industry has received negative publicity because of bovine spongiform encephalopathy (BSE) or mad cow disease. Because BSE can be transmitted to other animals by consumption of brain or spinal cords of infected animals, it is illegal to feed cattle animal products that may transmit the organism for mad cow disease. As a result many rendering operations refuse to accept cattle carcasses of animals that die on the farm. Thus many cattle producers have been forced to find safe, publically acceptable on-farm methods for carcass disposal.

Consumers can verify the origin of beef products by checking the product for country of origin labels (COOL) and mandatory identification procedures. Regulations for COOL and source verification increase management costs for producers and packers. However, source verification allows to the USDA to trace the offending producers in the event that contaminated or unsafe products reach supermarket shelves.

Another issue facing the beef industry is customer demand for product quality and consistency. See figure 9-30. Some beef industry personnel believe that increases in poultry consumption is due in part to the quality and consistency of poultry products. The beef industry has responded by "grid pricing" systems that reward producers for high-quality, high-cutability carcasses and providing feedback to producers on carcass cutouts. The goal is to assist the beef industry in delivering consistent, high-quality beef to consumers. Furthermore, many beef processors pursue added value through organic, all-natural, or other programs to differentiate their beef.

FIGURE 9-30

High quality, consistent beef cuts keep consumers coming back for more.

CAREER OPPORTUNITIES

Career opportunities in the beef industry begin with the producer. Large beef farms and ranches employ a variety of individuals. Managers and/or supervisors oversee day-to-day operations and direct the activities of farm/ranch laborers. Other farm/ranch employees may maintain records or work specifically with herd genetics. Personnel in feed and equipment businesses play supporting roles within the beef industry. In addition, educational consultants and veterinarians assist farmers/ranchers in efficiently producing beef. See figure 9-31. Once the animal is ready for slaughter, other industry workers move the beef through processing to the consumer's table. Livestock buyers and sellers arrange for transfer of animals among the types of operations and ultimately to the packer for slaughter. Many positions are available to individuals interested in the packing industry. These jobs include federal inspection, beef grading, quality assurance, and quality control positions. After the slaughter process has been completed, packaged beef is transported to retail markets, such as grocery stores, hotels, and restaurants. Salespeople working for packers initiate this transaction.

Besides producers, packers, marketers, and retailers, a variety of supporting individuals and companies

FIGURE 9-31

Beef industry jobs span the range from ranch hands to high-tech reproductive physiology

facilitate the efficient production and sale of beef. Other supporting personnel include animal scientists, educators, and veterinarians.

SUPPORTING ORGANIZATIONS

Every time a beef animal is sold, $1 is collected by state beef organizations to fund the beef check-off. The National Cattlemen's Beef Association (NCBA), with offices in Colorado, Chicago, and Washington, DC, oversees the expenditures of beef check-off money for product promotion, research, and information. It also acts as a coordinator of state beef organizations.

On a state level, 45 of 50 states have beef councils supported in part by check-off dollars. These councils also serve a promotional function. In addition, many states also have cattlemen's associations, which act in advocacy capacities. In some states, the beef council and the state cattlemen's association are housed together.

The U.S. Beef Breeds Council serves as an umbrella organization to unite the many beef breed associations. Chairship of this group rotates among the associations. The U.S. Meat Export Organization also works to further beef and other meat exports. This organization is industry driven but is partially supported by government funding.

SUMMARY

Cattle and calves ring up cash receipts accounting for 21 percent of total agricultural commodity income. Beef animals are raised across the country with heavy concentrations of cow-calf operations in the Southeast and west, and feedlots in the Plains States. The first beef breed imported to the United States was the Longhorn. Shorthorns, Herefords, and Angus, all from the British Isles, arrived in this country in the late 1700s and 1800s. These three breeds greatly enhanced the carcasses of commercial cattle. In the 1970s, several continental breeds were imported. These relatively new imports have become increasingly popular.

Because of the availability of pasture and range grasses, beef producers often breed their cows to calve in spring. First-calf heifers frequently take longer to rebreed; hence producers often schedule them to calve a month earlier than mature cows. A 60-day breeding season usually follows the calving period. Commercial products can prevent and alleviate infestation of internal and external parasites. In addition, a herd health program can help ensure healthy cattle. Feeding programs differ according to sex, age, and purpose of cattle. Beef cattle require very little shelter. A windbreak to escape drafty conditions is all that is necessary. Paying attention to local, cash, and futures markets will assist producers in effectively marketing cattle.

The beef industry can be segmented into four types of operations: seedstock, cow-calf, grower, and finisher. Issues facing the beef industry include concerns for food safety, delivery of consistent quality and products, and source verification. Jobs within the industry span from ranch managers to food scientists. Several groups, organized nationally and statewide, provide support for the beef industry.

CHAPTER REVIEW

EXPERIENTIAL LEARNING OPPORTUNITIES

1. Internship positions are frequently available in the beef industry. Investigate internship possibilities with your state beef council or association.

2. Beef breeding projects are popular but time-consuming. However, FFA Supervised Agricultural Experience program participants are eligible to compete for proficiency awards in beef production.

3. Contact the National Beef Board (http://www.beefboard.org) about the Masters of Beef Advocacy (MBA) program. Learners in this online course are instructed to become effective spokespersons for the beef industry.

DEFINE ANY TEN KEY TERMS

backgrounding	eared cattle	performance testing
baldies	feeder calf	polled
brood cow	finisher	progeny
commercial cattle	finishing cattle	replacement heifers
compensatory gain	grower	settle
continental breed	intensive rotational	shipping fever
cow-calf	grazing	terminal sire
dressing percentage	lactating	weaning
dystocia	marbled	winter

QUESTIONS AND PROBLEMS FOR DISCUSSION

1. Which breed of beef cattle was introduced to America by the Spanish in the 1500s?

2. Name the three beef breeds that originated in the British Isles and dominated the American beef cattle industry until the mid-1900s.

3. Crossbred or purebred _____ (list breed) are referred to as eared cattle.

4. Importation of continental breeds mushroomed in this decade. Write the decade. _____

5. Beef calves are usually born during these two seasons. List the seasons. _____

6. List three causes of dystocia.

7. The breeding season should last about _____ days.

8. Write an equation showing the days required in a 370-day beef production cycle.

9. Describe the mid-gestation diet of a mature pregnant cow.

10. A producer who uses a rotational grazing system can raise a cow/calf pair on as little as _____ acres.

11. List two parasitic flies that lay their eggs in manure.

12. True or False. Ringworm is a worm that habitually circles.

13. Name the three respiratory diseases that fall into the shipping fever category.

14. Describe the only housing requirement for mature beef cows.

15. What is the current market price of finished steers?

16. Which state ranks first in beef cow numbers?

17. How many beef producers exist in the United States?

18. Most feedlots are centered in this region. Name the group of states.

19. List four segments of the beef industry.

20. Seedstock operations provide breeding stock to _____ producers.

21. What management practices occur within the first 24 hours after a calf is born?

22. Finished beef cattle weigh in the _____ range.

23. True or False. *E. coli* poisoning can cause death.

24. Each time a beef animal is sold, $_____ is collected for the beef check-off.

25. Beef calves are normally weaned when weighing about _____ pounds, at six to eight months of age.

26. Who oversees the handling of beef check-off dollars?

27. List three uses of beef check-off dollars.

28. Grid pricing rewards _____, high-cutability carcasses.

29. Which organization unites beef breed associations?

30. How many states have beef organizations?

10

Dairy Cattle Management and the Dairy Industry

FROM MOO TO YOU

Objectives

After completing this chapter, students should be able to:

▶ Give an overview of the dairy industry in the United States

▶ Categorize the physical and production characteristics of the six dairy breeds

▶ Describe primary and secondary signs of estrus (heat) and identify the ideal time to artificially inseminate a cow

▶ List the steps in the milking process

▶ List two reasons why feed intake is essential to a high-producing dairy cow

▶ Draw a lactation curve

▶ List three practices that prevent mastitis

▶ Discuss the housing requirements for dairy cattle

▶ Identify three services offered through the Dairy Herd Improvement Association (DHIA)

▶ Describe how most milk is marketed in the United States

▶ Explain the organization of the dairy industry

▶ Discuss current issues in the dairy industry

▶ Describe career opportunities in the dairy industry

▶ Name two organizations that play supporting roles in the dairy industry

Key Terms

body condition	dry cows	milk fever
cooperatives	free-stall housing	parlor system
cwt.	grade	placenta
Dairy Herd Improvement Association (DHIA)	ketosis	price floor
	mastitis	somatic cell count
displaced abomasum	metritis	stanchion

Career Focus

Anil was a dairy science major. He liked cows and was very interested in the fascinating process of feeding highly productive lactating cows. Anil did an undergraduate research project in a rumenology laboratory, studying how ruminants digest several untraditional feedstuffs. After graduation, he used this experience as a stepping stone to a position with a seed company that bred highly digestible corn hybrids for corn silage production. Today, Anil does research on new varieties of silage corn and explains the nutritional advantages of these varieties to dairy nutritionists in an effort to increase his company's sales.

INTRODUCTION

Of all the species of livestock, lactating dairy cattle are perhaps the most difficult to manage. High-producing dairy cows are bred to give tremendous amounts of milk that, without pinpoint management, can overwhelm the physiology of the animal. Dairy producers constantly struggle with the complexities of breeding, feeding, and managing an animal pushed to its biological limit.

The dairy industry may also be the most labor intensive of all livestock industries. Cattle require milking two to three times per day, seven days a week. Net farm income generated from milk sales is second only to that generated by grain and oilseeds. The value

FIGURE 10-1

Maximum feed intake is a key to high-producing dairy cows.

of dairy products sold nationally in 2007 exceeded $37 billion. California ranks first in dairy production. Like many other agricultural industries, the dairy industry is changing. Fewer dairy producers own more cows, which produce more milk per farm. As health-conscious consumers demand lower fat diets, dairy food scientists have responded with a variety of specialty products. Fat-free yogurt, cream cheese, and frozen dairy desserts now line our grocery shelves.

INDUSTRY OVERVIEW

California leads Wisconsin and ranks first in number of dairy cows, followed by New York, Pennsylvania, and Idaho. Large farms in the west and Southwest have changed the demographics of the dairy industry. Although traditional dairy states such as Wisconsin, Pennsylvania, and New York remain major players, large dairies in California, Idaho, and other western states often house many more cows per operation. For example, about 50 percent of California's dairies maintain more than 500 cows, compared to only 2 percent of Wisconsin dairies. Forty years ago, 2 million dairy farms operated in the United States. In 2008, fewer than 70,000 remained operational. While total cow numbers have slowly declined, average production per cow has increased to over 20,000 pounds of milk per cow each year. In fact, through improved genetics, nutrition, and management, milk produced per cow per year increased by 18 percent from 1998 to 2007.

In 2009, over 9.2 million dairy cows in the United States produced over 185 billion pounds of milk worth over $37 billion. The United States leads the world both in milk production per cow and in total milk production.

BREEDS

In the not-so-distant past, cattle were expected to provide meat, draft power, and milk for their caretakers. Over time, a few select breeds that excelled in milk production were established as dairy breeds.

HOLSTEIN

As recently as the 1960s, significant populations of six different dairy breeds existed in the United States. However, since then, Holsteins have dominated the industry. Currently, Holsteins make up well over 90 percent of the dairy cattle in the United States. Holsteins are officially known as Holstein-Friesians and hail from the Netherlands and northern Germany. Holsteins first arrived in the United States in the mid-1800s.

Since 1970, Holsteins have made tremendous genetic progress due to rigorous selection. Producers have chosen bulls proven to transmit genes for higher milk production. This selection process is one reason that Holsteins produce significantly more milk per cow than the colored (non-Holstein) dairy breeds.

Most Holsteins are black and white. See figure 10-2. However, like Angus cattle, this breed sometimes harbors a recessive red gene. Both horned and naturally polled Holsteins are included in the registry. Registrations of Holstein calves in the United States totaled 348,033 in 2008, easily making them the most popular dairy breed.

Holsteins produce the most milk of all the dairy breeds. However, total solids percent (butterfat and protein percentages combined with minerals and lactose or milk sugar) are often lower than other breeds. Mature Holstein cows weigh 1,500 to 1,750 pounds.

Photo courtesy of the Holstein Association USA

FIGURE 10-2

Holsteins are the most popular breed of dairy cattle in the United States. Their black and white color pattern is unique to the breed.

JERSEY

On the other end of the size spectrum is the Jersey cow, weighing an average of just 1,000 pounds. See figure 10-3. Jerseys were developed on the island of Jersey, located off the coast of France. Jerseys were first imported in the early 1800s. A Jersey's coat color ranges from light tan to almost black. In 2008, Jersey registrations reached 94,774 ranking them a distant second in popularity.

Jerseys' popularity stems from their ability to efficiently convert feed to milk. This efficient feed conversion is due to the lower body-maintenance needs of the Jersey. Although the amount of milk produced by Jerseys is relatively low compared to Holsteins, the total solids content is among the highest of all breeds.

BROWN SWISS

The third most popular dairy breed is the Brown Swiss (registrations totaled 10,824 in 2008). See figure 10-4. As the name implies, the breed originated in Switzerland. Brown Swiss first came to the United States in the mid-1800s. They are normally brown to gray and compare to Holsteins in size. Brown Swiss are known for their ability to produce milk in hot climates. Milk production is second only to Holsteins. Total solids content of the milk ranks in the middle of all breeds.

Photo courtesy of the American Jersey Cattle Association

FIGURE 10-3

Smallest of the dairy breeds, Jerseys are known for the high percentage of butterfat in their milk.

Photo courtesy of the Brown Swiss Cattle Breeders' Association

FIGURE 10-4

Brown Swiss are a large-framed breed.

Photo courtesy of the Ayrshire Breeders' Association

FIGURE 10-5

Ayrshires originated in Scotland. Their coat color is white and deep or cherry red.

AYRSHIRE

Ayrshires are a smaller breed in terms of registrations (4,763 in 2008) and size (mature weight averages about 1,200 pounds). See figure 10-5. Ayrshires are a red and white breed, but the reds are darker than those of the Guernsey. Ayrshires originated in the Ayr district of Scotland and were imported in the early 1800s. Milk production of Ayrshires ranks midway among dairy breeds. Total solids are relatively low.

FIGURE 10-6

Guernseys have an orange-red and white color pattern.

GUERNSEY

The Guernsey was developed on the Island of Guernsey, which is also located off the coast of France. See figure 10-6. Guernseys were first imported to the United States in the early 1800s. Guernseys are a medium-sized red and white breed (often appearing orange and white). Guernseys are larger than Jerseys. Mature cows weigh about 1,100 pounds. On an average, Guernseys produce more milk than Jerseys, but Golden Guernsey milk is lower in total solids than Jersey milk. The deep yellow or golden color of Guernsey milk is due to the presence of beta carotene, a precursor to Vitamin A. The Guernsey breed registered 5,101 animals in 2008.

MILKING SHORTHORN

Milking Shorthorn is another dairy breed in the United States. See figure 10-7. Their registrations totaled 3,150 in 2008. Milking Shorthorns originated from the same base stock as beef Shorthorns and may be red, white, red and white, or roan.

RED AND WHITE

The Red and White Dairy Cattle Association records pedigrees and promotes red and white colored dairy cattle. See figure 10-8. Although this association

Photo courtesy of the American Milking Shorthorn Society

FIGURE 10-7

Milking Shorthorns can be red, white, red and white, or roan. They originated from the same seedstock as the Shorthorn beef breed.

Photo courtesy of the Red and White Dairy Cattle Association

FIGURE 10-8

The Red and White breed accepts any dairy animal that fits the color description.

maintains an open herd book, most cattle are genetically based in the red mutation of the Holstein breed. Registrations numbered 4,020 in 2008.

(Number of animals registered for each breed courtesy of the respective breed associations.)

Animal Science Facts

What constitutes milk? Mostly water. In fact whole milk averages almost 88 percent water. Other components include the following:

Fat	3.6%
Lactose (milk sugar)	4.7%
Protein	3.3%
Ash (Calcium and other minerals)	0.8%

These numbers may vary by breed. For example, Jersey milk averages over 5 percent fat. Lactose, protein, and ash are combined into a category called solids-not-fat (SNF).

BREEDING

Unlike other species of livestock, crossbreeding is uncommon in dairy cattle. Crossbred dairy cattle appear to show negative heterosis, or at best, no heterosis. In other words, there is no advantage in milk production or milk composition from crossbred dairy cows. However, disease resistance and survivability may be improved with crossbred dairy cattle. Nonetheless, most dairy cows in the United States are purebreds or very high percentage **grade** (unregistered) animals.

Dairy producers were the first to adopt artificial insemination (AI) on a large scale. Most dairy calves are a result of AI, but some producers still keep a bull to ensure reproductive success. Because of greater semen volume and multiple services during heat, a bull can help in settling problem breeders. However, dairy bulls are typically ill-tempered and dangerous. Since few bulls are needed by the dairy industry, most bull calves are castrated and sold for veal or as feeder calves at less than a week of age. See figure 10-9.

Dairy producers using AI release cows onto an exercise lot (semi-confined pen with good footing) to watch for standing heat at least two times per day. See figure 10-10. Standing heat (standing still when another cow attempts to mount) is the primary sign that a

© anistidesign. Used under license from Shutterstock.com

FIGURE 10-9

Many dairy bull calves are castrated and fed for veal production.

FIGURE 10-10

These cows at a large California dairy are in an exercise lot where they can be observed for signs of heat.

cow is ready to conceive. However, several other signs may tip off an observant producer that a cow is near heat. Secondary signs of heat include nervous bawling, restlessness, attempts to mount other cows, clear mucous discharge from the vulva, and a sharp drop in milk production. Insemination should be timed 12 hours after the cow is first observed in heat.

THE MILKING PROCESS

Cows are normally milked twice per day although some producers milk three times per day. Three times per day milking increases production by about 15 percent but requires more labor. At milking time, the cow's teats are washed for two reasons: disinfecting the teats and triggering the release of oxytocin, which initiates milk let-down. The teats are then dried with an individual disposable paper towel. One inflation of the claw of the milking machine is placed on each teat, or quarter (the udder is divided into four quarters). See figure 10-11. A vacuum is applied to the inflation, which draws the milk from the udder. When milk flow stops, the claw is removed and each teat is dipped in an iodine solution to prevent bacterial invasion of the udder. Total milking time is about seven minutes.

FIGURE 10-11

One inflation of the milking machine claw is fitted to each of the cow's four teats.

DAIRY NUTRITION

Dairy calf nutrition starts at, or within, 24 hours of birth, since calves are generally weaned immediately after receiving colostrum. Cows are then returned to the milking herd after parturition (freshening), and calves are raised entirely by humans.

Dairy calves must have colostrum as soon as possible after birth. Unlike humans who receive antibodies from the mother while in the uterus, the dairy calf is born without antibodies in the bloodstream to protect it from disease. The small intestine of the calf is very porous for the first 24 hours after birth and readily absorbs antibodies from the colostrum. Frozen colostrum is often substituted if the calf is weaned before its first meal.

Six to eight pints of milk replacer should be fed to the calf daily for the first five to eight weeks of age. At one week of age, calves should have access to small amounts of high-quality grain calf starter. High-quality hay should be introduced at four weeks of age. Calves are usually weaned from the milk replacer when calf starter consumption reaches four pounds per day. The concentration of the milk replacer can be reduced for a week or two before it is replaced with clean water.

Calves are not born with a developed rumen capable of digesting forages. Newborn calves have a groove that transports milk from the esophagus, directly past the undeveloped rumen, and into the abomasum. See figure 10-12. The rumen slowly develops during the first 12 weeks of life as the calf begins to consume forages.

From 12 weeks to 1 year of age, heifers can be fed a grain mix containing a feed additive, such as monensin or lasalocid, and free-choice high-quality hay or silage. The additives improve feed efficiency by altering rumen function so that cattle can capture more energy from the feed they consume. Heifers should gain about 1.5 to 1.8 pounds per day during this period (Jerseys and Guernseys a little less). Heifers reach breeding size and puberty at 11 to 12 months of age. See figure 10-13 for sample diets for calves and growing heifers.

FIGURE 10-12

Calves have an esophageal groove that transports milk past the developing rumen into the abomasum.

Ingredient (lbs)	Calf Starter	Growing Heifers
Cracked, Shelled corn	1040	
Ground ear corn		1818
Oats or barley	400	
Soybean meal	400	150
Molasses	100	
Dicalcium phosphate	10	18
Ground limestone	30	
Vitamin and trace mineral salt	20	14
Total	2000	2000

1. Calves eating calf starter should be offered high-quality alfalfa hay free choice.
2. Seven hundred pound growing heifers should be six pounds per head per day if good-quality hay is offered free choice. The grain mix will change depending on the forages used.

FIGURE 10-13

Common calf starter and heifer diets.

After breeding, pregnant heifers should be fed free-choice, high-quality forage. Several pounds of grain mix may be combined with the forage, to ensure proper development and provide needed trace minerals and vitamins. Heifers from large breeds should weigh at least 1,200 pounds at 24 months of

Animal Science Facts

Many dairy feeds, particularly silages, contain a large amount of water. Corn silage may contain 65 to 70 percent water, whereas dry hay contains only about 10 percent water. To compare all feedstuffs on an equal playing field, feed amounts are calculated on a dry matter basis—the weight of the feed with all water removed. However, feed in an as-is form is measured on an as-fed basis. Conversion from as-fed to dry matter basis simply involves multiplying by the percent (in decimal form) dry matter. Diets are normally formulated on a dry matter basis and then converted to an as-fed basis before feeding.

Example:

Corn Silage = 70% water, 30% dry matter

50 pounds corn silage (as fed) × 0.3 = 15 pounds on a dry matter basis.

Reverse the formula to convert from dry matter basis to as-fed basis.

15 pounds dry matter/0.3 = 50 pounds corn silage as fed.

age, when they deliver their first calf. Smaller breeds should weigh 1,000 pounds or more. Care should be taken so that heifers do not become fat. Adipose tissue deposition in the udders of developing dairy heifers reduces lifetime milk production. See figure 10-13 for examples of calf starter and heifer diets.

Feeding lactating dairy cows is an art as well as a science. A brief guide for feeding follows. Feeding programs have two segments: the needs of the cow and the nutrients provided by the feed.

The nutritional needs of the cow are dependent on body size (minor factor) and milk production (major factor). The amount of milk produced is dependent on the stage of lactation. See figure 10-14. During early lactation (first 60 to 90 days after calving), milk production increases. Feed intake during early lactation may double from 25 to over 50 pounds of dry matter each day. Rebreeding also takes place during early lactation. Peak lactation occurs about two to three months after calving. High milk production increases the nutrient needs to a point where the cow cannot physically consume enough feed to meet her demands. Therefore, cows often lose body weight during peak lactation. This loss amounts to 200 to 250 pounds from calving to peak lactation. Peak lactation for Holsteins averages about 120 pounds or 15 gallons of milk per day. Some cows reach over 200 pounds of milk per day. Later-lactation is a period of rebuilding **body condition**, or muscle and fat cover. Cows are

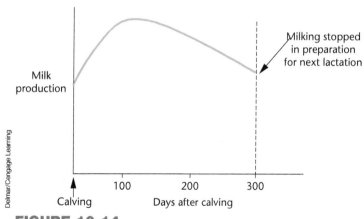

FIGURE 10-14

A lactation curve showing peak milk production approximately 100 days post calving.

"dried off" (milking is stopped) about 60 days before the next expected calving. Non-lactating cows are called **dry cows**.

Nutrients provided by feed depend on the quality and amount of feed intake. Large quantities of milk require large amounts of high-quality feed. As a rule of thumb, lactating dairy cows can be expected to eat 3.5 percent of their body weight on a dry matter basis. Three and one-half percent of a 1,500-pound cow equals about 52 pounds of feed on a dry matter basis. Two reasons for high levels of feed intake in lactating dairy cattle are: First, a pound of even the highest-quality feed can contain only a limited amount of nutrients; more pounds of feed consumed mean more nutrients available for milk production. Second, high feed intake is important for the maintenance of body condition. The metabolism of high-producing, early-lactating dairy cows prioritizes milk production over maintenance of body condition. Good body condition is critical for successful rebreeding. See figure 10-15. If nutrients are limited by low intake, cows will still give large quantities of milk, but body condition and reproductive success will suffer. High feed intake ensures enough nutrients for milk and maintenance of body condition. Techniques used to increase feed intake include multiple feedings per day and feeding forages before grain.

FIGURE 10-15

This cow is in good body condition for early lactation.

Delmar/Cengage Learning. Courtesy of Harvestore-Engineered Storage Products Co.

FIGURE 10-16

Forages are often stored in upright silos at a dairy operation.

Feeds used in lactating cow diets include forages. Forages may be dry hay, silage (fermented forages), or a mixture of hays and silages. See figure 10-16. Grain mixes may include any number of ingredients from standard corn and oats to ingredients such as beet pulp (by-product of the beet sugar manufacturing process) and cottonseed. Grain mix ingredients are used based on the specific nutrients needed and the price of available feedstuffs.

Forage quality influences the cow's ability to produce maximum milk at peak lactation. The amount of nutrients per pound of feed is critical. In addition, at least 50 percent of the diet should consist of forage to maintain normal rumen function. During peak milk production, the amount of forage in the diet can be reduced to 40 percent for several weeks; however, a continued diet at that level can lead to digestive problems. Obviously, the amount of nutrients contained in forages is very important to total nutrient intake. In addition, nutrients (particularly protein and energy) accounted for in the forage portion of the diet are generally less expensive than nutrients supplemented in the grain portion of the diet.

Grain mixes supplement forages based on the needs and milk production of each cow. For instance, a dairy producer may have access to a silo full of high-quality alfalfa haylage. A cow producing 40 pounds of milk per day in late lactation may need only 12 pounds of grain to supplement the haylage and meet her nutrient requirements. Another cow producing 110 pounds of milk per day at peak lactation may require 35 pounds of grain mix to supplement the same haylage.

Lactating cows are fed in groups according to the amount of milk produced. High-producing groups of cows receive the most grain. This high-producing group consists of cows in early and peak lactation. Alternately, low-producing groups (usually cows in later lactation) receive less grain. Computerized feeding systems on some farms allow cows access to grain depending on their individual milk production.

Vitamin and mineral supplementation for lactating dairy cows also varies depending on the forages and grains used. For example, cows fed a diet high in corn silage require more supplemental calcium than cows fed a diet containing large amounts of young alfalfa hay. (Young alfalfa hay contains much higher levels of calcium than corn silage.) Vitamins, minerals, and salt are normally added to the grain mix.

Other substances are often added to the grain mix of high-producing dairy cows. For example, buffers, such as sodium bicarbonate, are added to the diet of cows eating finely chopped corn silage or haylage. Buffers help keep the rumen at the ideal pH of 7.0, which helps maintain a healthy microbial population. Bypass, or rumen-protected fat (chemically coated fat particles that resist rumen degradation), is another feed additive. Cows in peak lactation are normally deficient in energy. Fat is a very high-energy feed ingredient, but the rumen is not equipped to process high levels of dietary fat. If too much rumen-available fat or oil is fed, it coats the forage particles, preventing the bacteria from digesting the forage. Rumen-protected fat bypasses rumen fermentation. Absorption takes place in the small intestine.

Many successful dairy producers employ a professional nutritional consultant, feed company nutritionist, veterinarian, or extension person to balance dairy diets. Because stored forage quality changes from the top of a silo or hayloft to the bottom, forages should be tested for nutrient content at least every 60 days. Diets can then be updated based on the nutrient values obtained from these forage tests.

Dry cows are normally fed a diet almost entirely composed of forages. However, they are often fed a small amount of grain mix to provide vitamins, minerals, and salt. Although body condition should recover during late lactation, the dry period can be used to improve body condition of thin cows. Grain intake should be gradually increased two to three weeks before calving to reacclimate rumen bacteria to large amounts of grain.

More information regarding cattle nutrition can be obtained from Cooperative Extension personnel or a publication called *Nutrient Requirements of Dairy Cattle* published by the National Research Council.

DAIRY CATTLE PARASITES AND DISEASES

Parasites and diseases of dairy cattle are much the same as those described for beef cattle in Chapter 9. A few exceptions are described here.

MASTITIS

Mastitis, an infection and inflammation of the mammary gland (udder), causes the greatest economic loss to the dairy industry. Mastitis can be acute or chronic. Acute (obvious or hot) mastitis is characterized by a hot, swollen udder and a noticeable drop in milk production. The milk may be bloody, clotted, or extremely watery. See figure 10-17. A cow with acute mastitis may be listless and may refuse to eat. Acute mastitis can be treated with high dosages of antibiotics. However, because milk from antibiotic-treated cows cannot be sold, producers may treat acute mastitis by injecting doses of oxytocin at hourly intervals followed by stripping the mastitic milk from the infected quarter. This practice reduces the number of bacteria in the udder and allows the cow's immune system to fight the infection.

Subtle, or chronic mastitis, is much less noticeable (subclinical). A few flaky appearing jets of milk at the beginning of milking may be the only sign. Chronic mastitis also causes a drop in milk production.

Prevention of either type of mastitis revolves around cleanliness. Controlling bacterial transfer through milking equipment, or while washing the udder in preparation for milking, is essential to controlling mastitis. See figure 10-18. Use of individual disposable paper towels to dry the udder after washing and before milking greatly reduces bacterial

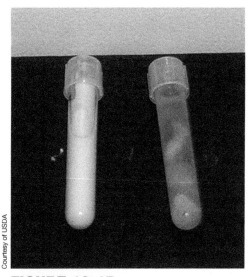

Courtesy of USDA

FIGURE 10-17

Mastitic milk is shown on the right compared to normal milk on the left.

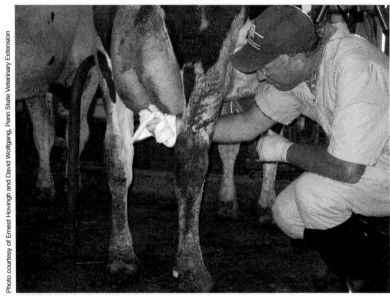

Photo courtesy of Ernest Hovingh and David Wolfgang, Penn State Veterinary Extension

FIGURE 10-18

Washing a dairy cow's udder before milking cleans the udder and hastens milk-let down.

transfer. Dipping teats in an iodine-based solution immediately after milking kills bacteria spread during the milking process. Disinfecting all milking equipment between milkings keeps the bacterial load at an acceptable level. Also, cleanliness of the cows' resting area helps keep udders clean between milkings.

Other health problems in the dairy industry result from the stress of high levels of milk production and the feeding programs required to reach those levels.

KETOSIS

Ketosis is a metabolic disorder of high-producing dairy cows associated with a negative energy balance during early lactation. Usually ketosis is not caused by underfeeding, but by a cow's sluggish appetite or being off feed due to dystocia, uterine infection, or digestive dysfunction. Cows with ketosis become listless and stop eating, and their breath smells of acetone. Treatment of ketosis often involves intravenous glucose injections. Cows with ketosis experience a marked decrease in total milk production.

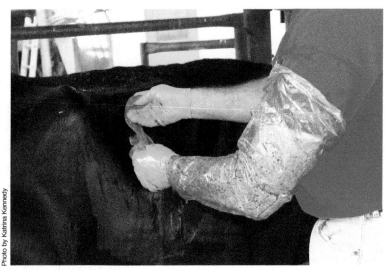

Photo by Katrina Kennedy

FIGURE 10-19

A veterinarian surgically repairing a displaced abomasum.

DISPLACED ABOMASUM

A **displaced abomasum** (DA) or "twisted stomach" occurs when the abomasum moves to an abnormal position in the body cavity. This condition is normally seen in very early lactation cows that are fed too much concentrate or silage before calving. The ejection of a calf leaves a large space in the cow's body cavity. The space allows the abomasum to swing from the right to the left side. Not more than 3 percent of a herd should have DAs in one year. A veterinarian should be consulted for treatment of a DA. See figure 10-19.

MILK FEVER

Milk fever is not really a fever but results from an imbalance of calcium in recently fresh cows. The imbalance causes muscle paralysis and prevents affected cows from standing. The extremities of cows with milk fever will feel cold to the touch. Calcium and phosphorus supplements during the dry period will aid in prevention. Treatment consists of a transfusion of calcium salts. Recoveries after treatment can be miraculous—sometimes within minutes. Not more than 5 percent of a herd should experience milk fever within a year.

RETAINED PLACENTA

Cows normally expel the **placenta** (membrane in the uterus that protects and nourishes the unborn animal) within hours after calving. If the placenta is not shed, a condition known as retained placenta occurs. Retained placentas quickly become infected and begin to decay within the uterus of the cow. Manual removal of the placenta by a veterinarian was once commonplace. Research has now shown that the weight of the placenta hanging from the vulva pressures the placental tissues to separate and is much less invasive. Retained placentas are blamed on a variety of causes including heat stress at calving, twinning, and low vitamin E and selenium in the bloodstream. No more than 7 percent of a herd should experience retained placentas in one year.

METRITIS

Retained placentas are one of several ways for bacteria to invade the uterus after calving. The resulting uterine infection is called **metritis**. Cows with metritis will exhibit abnormal discharge from the vulva, go off feed, and stand with their backs arched. Treatment involves infusion of antibiotics directly into the uterine body by a veterinarian.

DAIRY HOUSING

Housing for newborn and young dairy calves is often in individual stalls, either inside or preferably outside. Because of improved ventilation, calves housed outdoors have a lower incidence of respiratory disease than those raised indoors. Calf hutches are popular for housing calves after weaning. See figure 10-20. At eight weeks of age, heifers are normally grouped with other heifers of similar age. A separate heifer-growing barn is standard equipment for many dairy operations. Open front sheds are also popular heifer barns in many dairies.

Traditionally, dairy cows were housed in tie-stall or **stanchion** barns. In these facilities, cows were tied in individual stalls (stanchions) during milking and remained there for most of the day. Cows would be released into pasture at night during summer months but were kept in their stalls during the winter.

FIGURE 10-20

Young calves up to two months of age are often housed in calf hutches to help prevent the spread of disease.

Next came the advent of **free-stall housing**. See figure 10-21. Free stalls, as the name implies, allow cows to enter and leave the stall as they wish. A feed bunk is normally located in the center of the free-stall barn so that cows can eat at will. Cows housed in a free-stall arrangement are milked either in tie-stalls or in a milking parlor.

Milking in a tie-stall barn involves bending down to wash the cow's udder and bending again to attach and remove the milking machine. The person doing

FIGURE 10-21

In a free-stall housing system cows may enter or leave the stalls as they please.

Delmar/Cengage Learning. Courtesy of Catrina Kennedy and Scott Register

FIGURE 10-22

Milking parlors decrease milking time and increase labor efficiency.

the milking has to travel from one end of the barn to the other. This method involves many steps. The **parlor system** allows the cows to come to the milker. See figure 10-22. In a parlor system, a group of cows enter the parlor at one time. Positioning of the cows is such that the udders are at chest level for the milkers who stand in a pit behind the cows. All cows are washed and milked at the same time. When that group of cows is finished, they are released from the parlor, and the next group of cows enters. While parlors require a significantly higher investment, they dramatically increase the number of cows a person can milk per hour.

Robotic milking systems further reduce milking labor requirements. Robotic milkers allow cows access to milking 24 hours each day. When a cow feels the need to excrete milk, she enters the robotic milking stall. Sensors detect the cow's presence and, after mechanical washing, computer-controlled milking inflations adjust themselves to find the cow's teats.

RECORDS

Careful recordkeeping is key to any dairy enterprise. Records of milk production from individual cows alert producers to the best genetics in the herd. Production records also allow producers to cull lower-producing cows. Breeding records are important for keeping track of expected calving dates and keeping the average calving interval to an acceptable 14 months or less. With good records, feed costs and milk production can be monitored to determine profitability.

Dairy Herd Improvement Associations (DHIA) or similar recordkeeping organizations help producers keep track of important production records. Each month a DHIA tester weighs the milk produced and takes a milk sample from each cow. The samples are sent to a central laboratory where they are tested for protein, butterfat, and **somatic cell count** (an indication of mastitis and cleanliness). Somatic cells are white blood cells that fight infection. Key information about breeding dates, calving dates, and the feeding programs is recorded. Each month, the producer receives a report detailing the production of each cow, where she is in the reproductive cycle, herd production averages, and other production information.

MILK MARKETING

Most milk is either marketed through farmer **cooperatives** (members sharing in ownership and operation) or directly to processors. Producers in most dairy areas have a choice of two or more markets. Only those producers who pass a federal inspection for exceptional cleanliness and up-to-date facilities get to sell their milk as fluid milk used for drinking (Class I). See figure 10-23. Class I milk commands the highest price. Producers who do not meet requirements must sell their milk as Class II, which is used for processed milk products, such as butter and cheese.

Milk price is set by the federal government through Federal Milk Marketing Orders. The basis for this control is to keep milk prices stable throughout the year and at an acceptable level for the consumer. Producers have traditionally been paid bonuses

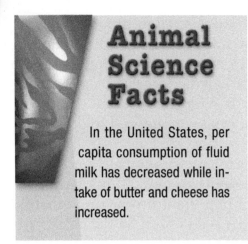

FIGURE 10-23

Only dairies that pass rigorous inspections for cleanliness can sell their milk as Class I for drinking.

for high levels of butterfat. However, consumers now want low and nonfat milk and milk products. Milk buyers have responded by paying an incentive for high solids-not-fat (protein). Bonuses are also paid for low somatic cell counts, giving producers an incentive to produce milk under clean conditions.

INDUSTRY ORGANIZATION

Dairy farms can be organized by the number of cows they maintain: less than 100, 100 to 500, over 500. The economies of scale enjoyed by larger operations make it difficult for smaller operations to compete financially. For example, inflated machinery prices force smaller dairies to continue with older, less-efficient equipment. Many small dairies of less than 80 milking cows survive only with second incomes generated from off-farm sources. Others hire custom farmers to do their fieldwork so the producers can concentrate on maximizing milk production.

Dairies with 100 to 500 milking cows can also find themselves in financially difficult situations. These producers need the same equipment as larger operations but have fewer cows to offset the additional equipment costs. Some small- and mid-size herds respond by implementing intensive rotational grazing, which

FIGURE 10-24

Grazing reduces costs by allowing the cows to harvest their own forage.

can reduce feed costs significantly. Pastured dairy operations may be managed seasonally so that all cows are dry during the winter when pastures do not grow. Intensive pasture management leads to lower veterinary costs due to improved foot health and lower mastitis occurrences. See figure 10-24.

Dairies with 500 or more milking cows have the financial advantages of operating on a large scale. They accounted for 52.5 percent of all dairy cows in the United States in 2007. While smaller dairies are often family-run businesses, large-scale dairies may be corporately owned, even if the corporation consists of extended family. These businesses often employ professional managers to oversee specific divisions, such as crops or cows. Furthermore, some large farms are taking advantages of economical and non-traditional feedstuffs, such as citrus pulp and soybean hulls. See figure 10-25.

INDUSTRY ISSUES

Profitability tops the list of dairy industry concerns. United States' dairy farmers routinely produce a milk surplus. Government programs guarantee a milk **price floor** (lowest possible price). The government price floors are established in the U.S. Farm Bill by setting

Delmar/Cengage Learning. Photo by Frank Flanders

FIGURE 10-25

By-product feeds. Top row left to right: Citrus pulp, Whole cottonseed, Soybean hulls. Bottom row left to right: Corn meal, Distillers grains, Soybean meal.

price floor per **cwt**. (hundred pounds). Depending on the details of the current order, producers may receive a portion of the difference between the actual market price and the established floor price if market price falls below the floor price.

Wholesale fluid milk prices have vacillated drastically in recent years. When prices are high, well-managed dairy farms can be profitable. However, during periods when prices are low, producers are saddled with increased input costs, and profits can be nonexistent for long periods of time. The result of the profitability squeeze is an increase in cow numbers to take advantage of economies of scale. Larger farms can also amass enough capital in good times to survive the difficult periods. This shift from small to larger farms is well underway. Another factor working against smaller dairy farms is the inheritance tax. If proper estate planning is not done well in advance, the taxes owed upon the death of a dairy farmer can leave heirs with little choice but to sell the farm.

Another consequence of dairy consolidation is the loss of dairy industry infrastructure in areas where cow numbers are declining. Loss of supporting businesses makes dairying difficult.

Environmental issues also challenge the dairy industry. On farms where all forages and grains are grown locally, manure nutrients can usually be applied to the local land base and used for crop production. However, on larger dairies, grain is sometimes purchased from off-site. In this case, nutrients found in manure may be more than can be applied to local fields. See figure 10-26. In addition, dairy cows use quite a bit of water to produce milk. In some areas of the United States, citizens are concerned that well water used by local dairies may depress groundwater levels and threaten neighboring wells. Dust generation from large California dairies have also come under environmental scrutiny.

Finally, animal welfare concerns have been raised in the dairy industry. Some consumers are concerned that contemporary dairy cattle residing in large-scale production facilities do not receive appropriate care. Concerns are largely based on images of excessively thin cull dairy cows. Producers should take care to rehabilitate thin cows before marketing or euthanize them on the farm.

Photo courtesy of Robert Meinen

FIGURE 10-26

Manure application equipment is capable of injecting nutrients under the soil and thus reduce odors.

Photo by Peggy Greb. Courtesy of USDA

FIGURE 10-27

Dairy veterinarians are in short supply in some parts of the country.

CAREER OPPORTUNITIES

Most jobs in the dairy industry are located in sales. Feed, machinery, equipment, and technical sales and service personnel are needed to provide dairy farmers with supplies and services for efficient operation. For example, dairy semen can be purchased from breeding cooperatives who can send artificial insemination technicians to service cattle in heat. Additional job opportunities exist in cooperatives that purchase milk from farmers, process dairy products, and market them to retailers. Milk cooperative personnel include truckers, field service individuals, managers, inspectors, as well as processing plant operators, supervisors, and marketing specialists.

Potential dairy employees may also want to consider becoming DHIA workers who provide recordkeeping services to producers. Those interested in communications may wish to seek careers in breed associations or with dairy-oriented media and promotion organizations. Dairy food scientists constantly research new products that will appeal to health-conscious consumers. In addition, veterinarians provide valuable herd health services for dairy farmers, such as ration development, herd health maintenance, and disease prevention. See figure 10-27.

SUPPORTING ORGANIZATIONS

Several organizations support the dairy industry at the national level. The National Dairy Board and United Dairy Industry Association (UDIA) united in 1995 to form Dairy Management Incorporated, which is located in Illinois. This group represents the entire country's dairy industry. Dairy Management Incorporated promotes research, education, and the use of milk products. Dairy Management Incorporated supervises the expenditure of dairy check-off dollars. Like other commodity groups, the dairy check-off program began in 1984 as an effort to finance milk promotion. Today, check-off monies are collected at a rate of 15 cents cwt. of milk with 5 cents going to national efforts and 10 cents remaining with local agencies qualified by the USDA. Memorable slogans like, "Milk, it does a body good" and "Got milk?" are a result of dairy check-off dollars.

The National Milk Producers Federation (NMPF) consists of milk marketing cooperatives and independent dealers of fluid milk. This organization represents the interests of the milk marketing cooperatives and dealers to various segments of government and society. For example, NMPF works with USDA to develop the federal government's milk marketing regulations. The International Dairy Foods Association (IDFA), headquartered in Washington, DC, is an organization that represents the interests of the dairy foods processing industries around the United States. Member organizations of IDFA include, but are not be limited to, The Kroger Company, Kraft Foods, and Borden.

SUMMARY

Traditional breeds of dairy cattle in the United States are limited to only six: Holstein, Jersey, Guernsey, Brown Swiss, Ayrshire, and Milking Shorthorn. Holsteins are by far the most popular. Most dairy cows are artificially inseminated. Crossbreeding is rarely practiced in the dairy industry. The milking process is designed to be easy on the milker, sanitary, and comfortable for the cow. Milking can be done two or three times per day. Dairy calves are fed milk or milk replacers after weaning, which occurs shortly after birth. Growing heifers should be fed limited amounts of grain and high-quality forages to ensure proper growth and development. Lactating cows are fed larger amounts of grain, but diets always contain at least 40 percent forage. Forage quality and body condition are very important components in feeding the lactating cow. Cows are fed more grain in early and mid-lactation period when milk production is highest and less in later lactation period. Metabolic disorders are common in dairy cows because of the high demands of lactation. Mastitis is the most costly of all dairy cow ailments. Most cows are housed in either tie-stall barns or free-stall barns. Milking is normally done in tie-stalls or in a milking parlor. Production records, such as those provided by DHIA, are critical to profitability. Milk price is ultimately set by the federal government.

Like most agricultural commodities, the dairy industry is changing. Large dairies have dramatically affected the makeup of the industry. Fewer dairies, each with more cows, now exist. Dairy farms can be organized into three groups: 100 or fewer cows, 100 to 500, or more than 500. Inconsistent profitability plagues the dairy industry, prompting more large-scale dairies. Input costs continue to rise. Estate taxes, loss of infrastructure, environmental issues, and animal welfare also strive for the attention of dairy farmers. While many job opportunities exist in the dairy field, sales careers are most prevalent. Several associations play supporting roles in the dairy industry.

CHAPTER REVIEW

EXPERIENTIAL LEARNING OPPORTUNITIES

1. Participate in the FFA Dairy Foods career development event. Enlist the help of a dairy case manager from a local grocery store.

2. Develop a research project based on the ability of participants to differentiate among fluid milk with varying fat content. Conduct the trial and exhibit the results in the FFA Agriscience competition.

DEFINE ANY TEN KEY TERMS

body condition

cooperatives

cwt.

Dairy Herd Improvement Association (DHIA)

displaced abomasum

dry cows

free-stall housing

grade

ketosis

mastitis

metritis

milk fever

parlor system

placenta

price floor

somatic cell count

stanchion

QUESTIONS AND PROBLEMS FOR DISCUSSION

1. Which dairy breed is the most popular in the United States?

2. Jerseys were first imported to the United States in the early _____ (date).

3. Name the primary sign of estrus (heat).

4. List three secondary signs of estrus (heat).

5. To achieve the best rate of conception, insemination of cows should occur _____ hours after the first observation of standing heat.

6. Describe the two main purposes for washing a cow's udder prior to milking.

7. List the two major categories of feedstuffs in a lactating dairy cow's diet.

8. Lactating dairy cows can be expected to consume _____ percent of their body weight on a dry matter basis.

9. Cows are often fed in groups according to _____.

10. Draw a lactation curve.

11. Which health problem in dairy cattle causes the greatest economic loss?

12. What causes milk fever?

13. The primary use for Class I milk is _____.

14. Differentiate between free-stall and stanchion barns.

15. Most milk in the United States is marketed through _____.

16. Which state ranks number one in dairy production?

17. Forty years ago, over _____ million dairy farms operated in the United States.

18. The average cow produces over _____ pounds of milk per year.

19. Explain how dairies can be segmented into three groups by number of milking cows maintained.

20. How can intensively managed pastures cut veterinary bills?

21. Name one nontraditional feedstuff used by some large California dairies.

22. What does cwt. mean?

23. True or False. Milk prices have been very stable in recent years.

24. True or False. Dairy producers should send extremely thin cows to market.

25. The biggest area of employment in the dairy industry is _____.

26. The dairy check-off program began in _____.

27. The _____ is comprised of milk marketing cooperatives and independent dealers of fluid milk.

28. Name the location of The International Dairy Food Association.

29. Which group provides recordkeeping services for diary producers?

30. Dry cows are mostly fed _____.

11

Sheep Management and the Sheep Industry

BAA BAA BLACK SHEEP

Objectives

After completing this chapter, students should be able to:

▶ Give an overview of the sheep industry in the United States

▶ Discuss the characteristics of the major sheep breeds

▶ Explain the breeding systems used in the sheep industry

▶ Discuss nutrient requirements for various stages of sheep production

▶ List five management practices used in the sheep industry

▶ Describe common sheep diseases

▶ Make a chart of preventive health care measures used in the sheep industry

▶ Discuss the housing requirements for sheep

▶ Inventory various marketing channels for lamb and wool

▶ Differentiate between farm and range flocks

▶ Explain current issues and trends in the sheep industry

▶ List three careers in the sheep industry

▶ Name two organizations that play supporting roles in the sheep industry

Key Terms

clip	flushing	scur
creep feeding	hothouse lamb	seasonal breeder
crutched	jockeys	stillborn
docked	medium-wool breed	synthetic breed
drench	niche market	
fine-wool breed	predators	

Career Focus

When Penny was 12 years old, her grandmother taught her the art of knitting. Penny loved to knit shawls, sweaters, and socks, and her favorite yarn was wool. She loved the warmth and heft of the finished woolen garments. When Penny was a teenager, she became interested in yarns from different types of sheep. She experimented with fine and coarse wools, and began to track the type of sheep from which the different wools came. She even purchased a small flock of fine-wool sheep and experimented with cleaning, carding, and spinning her own wool yarn. As Penny's knitting skills improved, she found a ready retail market for her woolen clothing. Penny attended a two-year business course after high school and used her business skills to develop a small thriving company that sold hand-knitted woolen garments.

INTRODUCTION

The history of the American sheep industry is clouded in conflict. For years, shepherds and cattle ranchers on the western range were at odds over whose animals would graze the available land. Both sheep and beef cattle convert forages into meat, but sheep take production one step further by providing an annual yield of wool. The feud has since cooled, but sheep still excel at converting forages into both food and fiber.

Compared to other livestock species, initial setup in the sheep industry is relatively inexpensive. Sheep utilize forages and move from conception to a salable market lamb in eight to nine months. Furthermore, sheep produce both wool and meat for human consumption. Still, sheep inventories have dropped by over two-thirds during the last 30 years. In 2007, the U.S. Department of Agriculture stated a national inventory of 5.8 million sheep. Moreover, average consumption of lamb and mutton continues at less than 1 pound annually, significantly lower than the seven-pound annual average in the early 1940s. Sheep populations are concentrated in the western states where larger flocks graze rangeland. Smaller eastern farm flocks supply lambs to the East Coast

FIGURE 11-1

These lambs will grow quickly by eating this excellent quality grass.

metropolitan market. General decline tops the list of sheep industry issues. The American Sheep Industry Association speaks on behalf of the sheep producers and industry personnel.

SHEEP INDUSTRY OVERVIEW

The western range states account for much of the U.S. sheep population. Even so, sheep often take a backseat to other agricultural commodities. For example, Texas ranks first in the nation for sheep cash receipts ($107 million), but sheep sales account for only 0.7 percent of the livestock production value in the state. The top five states in sheep numbers are Texas (945,000 head), California (596,000 head), Colorado (413,000 head), Wyoming (412,000 head), and South Dakota (335,000 head). Wool value is not included in sheep cash receipts. In 2007, 34.5 million pounds of wool was shorn with an average market price of $0.88 per pound. Total value of wool production in United States was $30.3 million in 2007.

(USDA 2007 Census of Agriculture
Canada, Carol. 2008. Wool and mohair price support.
Congressional Research Service Report for Congress.
RS20896)

BREEDS

The sheep industry in the United States consists of nearly 50 different breeds of sheep. Of these, only nine have significant registrations. Breeds of sheep can be classified by place of origin, purpose (meat versus wool production or sire versus dam breeds), and type of wool.

SUFFOLK

According to registrations, the most popular breed in the United States is the Suffolk. See figure 11-2. Suffolks were developed in the late 1700s in a region of southern England called the Downs. Suffolks, along with other breeds originating in that region, are known as the Down breeds that were developed for meat production and are often used as sire breeds. Large-framed and heavily muscled, Suffolks sire fast-growing market lambs.

Suffolks and the other Down breeds are referred to as **medium-wool breeds** that have average fleece quality. The head and legs of Suffolks are black and

Photo courtesy of Sheep Breeder Magazine

FIGURE 11-2

Suffolks are the most popular breed of sheep.

covered with hair instead of wool. Their ears are long and somewhat drooping. The only evidence of horns is an occasional **scur**, or small, unattached horn. Suffolks were first brought to the United States in the late 1800s; however, most imports occurred during the twentieth century. In 2008, 11,034 Suffolks were registered in the United States.

DORSET

The second most popular breed in the United States is the Dorset (approximately 9,500 registrations in 2008). See figure 11-3. Dorsets are also a medium-wool breed, but not a Down breed. They were selectively bred from native sheep in a small area located outside the Downs region. The average Dorset is smaller but heavily muscled like a Suffolk. Dorset ewes are popular in crossbreeding systems because of their ability to breed throughout the year. Most sheep are **seasonal breeders** (able to conceive only in the fall of the year). Because of their muscling and maternal abilities, Dorsets can be used as either a sire or a dam breed.

Photo courtesy of Sheep Breeder Magazine

FIGURE 11-3

Dorsets are white-faced sheep used as both a sire and dam breed.

They are entirely white, open faced (free from wool around the eyes), and have short ears. Dorsets can be either horned, polled, or scurred. They were first imported to the United States in the late 1800s.

HAMPSHIRE

Ranking a close third in popularity with 7,584 sheep registered in 2008 are Hampshires. See figure 11-4. A medium wool breed, they are also heavily muscled. Like Suffolks, Hampshires developed in the Down region. They sire fast-gaining, heavily muscled market lambs and are one of the larger Down breeds. They have a black face and legs, short ears, and are polled. The top of a Hampshire's head is covered with wool. Hampshires were also brought to the United States in the late 1800s.

RAMBOUILLET

The Rambouillet, a **fine-wool breed** (excellent wool quality), can trace its ancestry to the Spanish Merino breed (from France), which was known for superior wool quality. See figure 11-5. Since importation to the United States in 1840, Rambouillets have been selected for improvement of carcass qualities. They are popular with western sheep producers as a dam breed due to their ability to wean lambs with acceptable carcass merit and to produce a high-quality, heavy

Photo courtesy of Sheep Breeder Magazine

FIGURE 11-4

Hampshires compete with Suffolks for use as market ram sires.

Photo courtesy of Linde's Livestock Photos

FIGURE 11-5

Rambouillets are a fine-wool breed.

fleece. Rambouillets are a large-framed, white-faced breed that can be either polled or horned. There were 3,008 Rambouillets registered in 2008.

POLYPAY

In 2008, 1,262 Polypay sheep were registered in the United States. See figure 11-6. Polypays are a **synthetic breed** (new breed from planned matings) developed in the late 1960s and early 1970s. They accentuate the various strengths of four breeds, the Dorset, the Finnsheep, the Rambouillet, and the Targhee. The resulting Polypay has excellent maternal abilities, high growth rate, good carcass qualities, and hardiness. They are often used as a dam breed because of their ability to conceive in all seasons. Polypays are white, medium-sized, are open faced, and have medium-sized ears.

COLUMBIA

Another synthetic breed developed in the United States is the Columbia. See figure 11-7. Columbias were first produced in the early 1900s from a foundation of Lincoln and Rambouillet breeding stock. They are a white-faced breed with open eye channels, somewhat resembling the Rambouillet. Columbias are popular in western range operations, where they are often crossed with Suffolks or Hampshires. Registrations totaled 2,009 head in 2008.

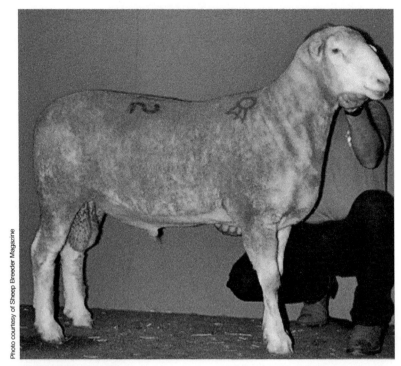

Photo courtesy of Sheep Breeder Magazine

FIGURE 11-6

Polypays are utilized as a maternal breed.

Photo courtesy of the Columbia Sheep Breeders Association

FIGURE 11-7

Columbia ewes populate many western ranges.

FIGURE 11-8

Although small in size, Southdowns are known for their muscle volume.

SOUTHDOWN

One of the first breeds of sheep to be introduced to the United States is the Southdown. See figure 11-8. They were fourth in popularity in 2008 with 5,222 registrations. Easily the smallest of the Down breeds, Southdowns are extremely heavily muscled. Their face and legs are covered with brownish wool. Southdowns are popular on the East Coast for producing early-maturing, heavily muscled, grain-fed lambs for the hotel, restaurant, and ethnic markets.

CORRIEDALE

Corriedales were developed in New Zealand in the late 1800s from Merinos, Lincolns, and Leicesters. See figure 11-9. Because of their similar genetic heritage, the appearance of Corriedales is similar to that of Columbias. The idea behind such crosses was to reproduce the fine-wool qualities of the Merino and the carcass qualities of the Lincolns and Leicesters. Corriedales were first imported into the western United States in the early 1900s. They have been used as a dam breed in range flocks. Registrations numbered 1,522 in 2008.

FIGURE 11-9

Corriedales originated in New Zealand.

Photo courtesy of Linde's Livestock Photos

FIGURE 11-10

Shropshires are a down breed that can function in either a sire or a dam role.

SHROPSHIRE

Shropshires are a Down breed that can function in either a sire or a dam role. See figure 11-10. They have brownish-black pigmented faces, ears, and feet. Shropshires are medium-sized sheep, ideal for eastern farm flocks. They were first imported to the United States in the late 1800s. There were 2,333 Shropshires registered in 2008.

TEXEL

Texel sheep are a relatively new addition to the U.S. sheep industry. See figure 11-11. Texels are extremely heavily muscled, lean, and serve as a terminal sire breed. Texels were developed on the Isle of Texel near the coast of the Netherlands. The Texel Sheep Breeders Society registered 333 animals in 2008, although Texels and Texel crossbreds have enjoyed wide commercial acceptance.

(Registration data courtesy of the respective breed associations.)

Photo courtesy Richardson Texels/Texel Sheep Breeders Society

FIGURE 11-11

Texels are the heaviest-muscled sheep breed.

BREEDING SYSTEMS

Sheep breeds and breeding systems vary greatly throughout the United States. The western plains and Rocky Mountain states are collectively known as the

Photo courtesy USDA

FIGURE 11-12

Western range flocks may spend summers in high mountain meadows.

western range. Flocks in western range country often number in the thousands. See figure 11-12. Breeding systems on the western range rely on large, hardy, white-faced ewes with better than average wool production. Breeds such as the Rambouillet, Columbia, and Corriedale are popular as ewes in the western ranges. Wool production has traditionally been a major source of income for western sheep producers, hence the emphasis on ewes with a heritage of high-quality wool. The other source of income is from market lambs. Western range ewes are routinely bred to heavily muscled, growthy breeds, such as Hampshires or Suffolks to produce market lambs with improved growth rates and carcass qualities. Replacement ewes are either purchased or produced by mating a maternal, fine-wool ram with a band (group) of ewes and retaining the female offspring.

Ewes are shorn (wool removed) and lambed in the spring. Ewes and lambs are pastured during the summer; sometimes in high mountain meadows, and often on public grazing lands. In late summer, the lambs are weaned and the rams are turned in with the ewes. The rams remain with the ewes for at least three estrus periods, or 48 days. Well-grown, eight-month-old ewe lambs can be bred the first year. However, better reproductive performance results if ewe lambs mature to one and one-half years of age before breeding. When summer grass conditions are adequate, weaned lambs are sold for slaughter. Lambs deemed too thin for slaughter are often sold or transported to feedlots for finishing.

Sheep flocks in the East and Midwest are much smaller than their western counterparts. They usually number fewer than 200. Their breeding programs are also different. Because of the proximity to populated areas and large quantities of grain, the majority of income is derived from the sale of market lambs. See figure 11-13. Ethnic groups concentrated in large cities comprise a major market for lamb. Therefore, Eastern and Midwestern producers have concentrated on meat breeds, such as the Suffolk, Hampshire, Dorset, and Southdown. Progressive producers primarily use maternal breeds (Dorset, Polypay, or crosses) that conceive during any season. Ewes of these breeds produce a lamb crop every 8 months instead of every 12 months. As with western producers, black-faced

FIGURE 11-13
Market-ready lambs in a feedlot.

rams are used as market lamb sires. Lambs born "out of season" (in the fall or winter) can be marketed as **hothouse lambs** (raised indoors) at 9 to 16 weeks of age. Spring-born lambs are normally weaned at 70 to 80 pounds and finished on grain.

SHEEP NUTRITION

As with the beef cow herd, feeding programs for the ewe flock are almost entirely forage-based. Sheep are pastured during the grazing season and fed stored forages the rest of the year. Feeding programs for western range ewes managed for an annual lamb crop are based on a yearly production schedule. Summer pastures for western range flocks are composed almost entirely of native grasses. See figure 11-14. Shepherds move the flock to a new area when the grass supply is exhausted. A grazing area may be used twice during the grazing season if sufficient time is allowed for regrowth. When the grazing season ends, sheep are moved to a winter range. The winter range is normally at a lower elevation or in a more sheltered location, where the grazing season may be longer. When pasture runs out, ewe flocks are wintered on a variety of stored forages, such as mixed legume hay or silage. Root crops, such as turnips, can be used as an early winter feed.

Photo courtesy USDA

FIGURE 11-14

Sheep make good use of native forages, turning them into high-quality meat and wool.

After the lamb crop is weaned, the ewes are prepared for the breeding season by **flushing**. Flushing consists of increasing the energy content of the diet for 15 to 20 days before breeding. A ration of 1/2 to 1 pound per head per day of oats, corn, or a similar grain, in addition to pasture or stored forage, is recommended. This practice increases the number of eggs ovulated and, hence, the number of lambs born per ewe (lambing percentage). See Animal Science Fact in this chapter.

Nonlactating, pregnant ewes in early gestation have lower nutritional requirements than at any other time during the production cycle. In some parts of the United States, crop residue or wheat pasture is used as the sole feed. Free-choice legume or legume mixed hay is probably the most popular stored forage when pasture is unavailable.

Later in the 150-day gestation period, the rapidly growing fetuses require additional nutrients. Ewes also need to build body condition before the next lactation. About five weeks before lambing, ewes are given supplemental grain to complement the available forage. For example, if ewes are being wintered on low-protein forage, 1/2 pound per head per day of a high-protein grain mix may be in order. Thin ewes receiving adequate protein from forage may simply require 1/2 to 1 pound of high-energy grain per day.

Decisions on the exact supplement should be based on the forage quality and body condition of the ewe.

Lactation places great nutrient demands on the ewe. See figure 11-15. Nursing lambs, wool production, body growth for young ewes, and the maintenance requirements of the ewe herself require the best forages available. The best quality legume hay should be reserved if lactating ewes are fed stored forages. Free-choice hay and 1 to 2 pounds of high-energy grain mix are necessary. See figure 11-16 for an appropriate grain mix. Spring-born ewes and lambs

Photo courtesy USDA

FIGURE 11-15

Lactation places high nutritional demands on a ewe.

Ingredient (lbs)	Lactating Ewes	Creep	Finishing Lambs
Cracked, shelled corn	60	50	81
Oats or barley	30	30	
Soybean meal	10	15	16
Molasses		5	3
Total	100	100	100

1. Salt, vitamins, and minerals *manufactured specifically for sheep* should be offered free choice to all ewes. Vitamin/mineral premix *manufactured specifically for sheep* must be added to creep feeds and finishing diets in accordance to manufacturers' recommendations.

2. Good-quality legume or legume mix hay should be offered free choice for lactating ewes, creep-fed lambs, and finishing lambs.

Delmar/Cengage Learning

FIGURE 11-16

Rations for lactating ewes, creep-fed, and finishing lambs.

should be released to high-quality pasture as soon as possible and the supplemental grain feeding should be eliminated. Many managers separate ewes that have twins or triplets from those that have single lambs. The more productive groups of ewes can then be fed for maximum milk production. Ewes with single lambs can often go without supplemental grain.

Eastern farm flocks also utilize pasture when in season. Managing ewes for an eight-month production cycle means some ewes will lactate while being fed stored forage in the winter. Other dry, pregnant ewes will have access to lush spring and summer pastures. For these reasons, feeding ewes on an eight-month lambing cycle requires more nutritional management than feeding those on a 12-month cycle. The nutrition of ewes should be managed so that those with the highest demands (late gestation and lactation) receive the best pasture or stored forage, and grain. Dry, pregnant ewes in early gestation can be fed poorer-quality pasture or stored forage.

Nutrition in the eastern flock starts with **creep feeding** (allowing lambs access to special, high-quality feed) before weaning. See figure 11-17. Creep feed should be available free-choice, in an area where only lambs have access. Creep feeding serves two purposes. First, it acclimates lambs to eating grain.

Delmar/Cengage Learning

FIGURE 11-17

This lamb is consuming a high-quality creep feed.

Second, it increases weaning weights. Ewes nursing twins and triplets may not supply enough milk to fully satisfy each lamb. Creep feed can fill this milk shortage. Rolled oats, crimped corn, soybean meal, molasses, and other palatable feedstuffs are common creep feed ingredients. See figure 11-16.

When lactating ewes have access to good pasture, lambs weaned at five or six months are often fat enough for slaughter. Weaned range lambs not ready for slaughter are normally sold to feedlots for two or three months of grain feeding. Eastern lambs that are not sold early as hothouse lambs are also fed a finishing diet for a two- to three-month period.

Finishing diets should contain at least 1 pound of high-quality legume forage per head per day. The protein and energy in such forages are normally less expensive than those in grains. Like finishing cattle, lambs require a small amount of forage to maintain normal rumen function. Grain mix ingredients are variable but should be high in energy. A popular finishing diet consists of whole or cracked grains mixed with a fortified, pelleted supplement. See figure 11-16. Lambs should be started on grain slowly. The initial feeding should be 1/4 pound per day. Feed allowance should be increased by 1/4 pound per day every three to five days until a maximum grain intake of 2 to 2 1/2 pounds per day is achieved.

Finishing lambs should gain at least 1/2 pound per day. Finished weight depends on genetics and the weight at which grain feeding begins. Most lambs require less than three months of grain feeding before slaughter.

Finished lambs do not require much exterior fat cover. Buyers want trim lambs in the 90 to 150 pound range depending on the market and lamb breed or cross. Fat cover should be sufficient to prevent excessive water weight loss (shrink) from the hanging carcass (at least 0.10 inch) but not enough to cause excessive trimming (less than 0.20 inch). See figure 11-18.

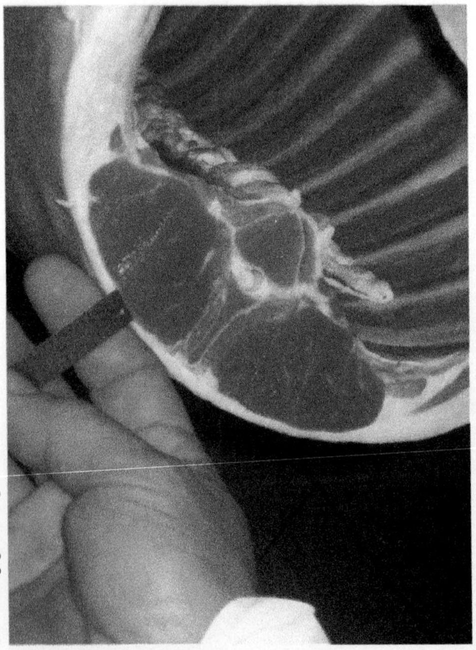

Delmar/Cengage Learning

FIGURE 11-18

This lamb carcass measures approximately 0.15 inches of backfat, ideal for a market lamb.

SHEEP MANAGEMENT

Sheep require several management practices not necessary in other species of livestock.

Tails are routinely **docked** (removed) in young lambs. See figure 11-19. Because sheep do not have the ability to lift their tails very high, undocked tails collect feces, which serve as an ideal place for maggots to grow. Tail docking should be completed within days of birth when lambs are easily handled. Tails may be cut with a hot pincher, which stops the wound from bleeding and helps prevent infection. Alternatively, tails can be banded with an elastrator, which cuts the blood supply to the tail and the tail eventually falls off.

Castrated wether lambs entering a feedlot can be implanted with a growth promotant. Implants are injected just under the skin on the backside of the ear. Implants increase average daily gain and improve feed efficiency. However, care should be exercised when using growth promotants as their use may cause some side effects, such as rectal prolapses.

Shearing is the removal of wool from the sheep. See figure 11-20. Ewes are normally shorn once per year before lambing. If lambing is extremely early when inclement weather is still possible, shearing may be delayed until after lambing. After removal, the wool is called a fleece. The average full fleece for a mature

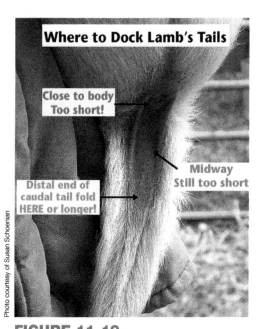

FIGURE 11-19

Tails are usually docked when lambs are very young.

FIGURE 11-20

Experienced shearers can remove a fleece in just a few minutes.

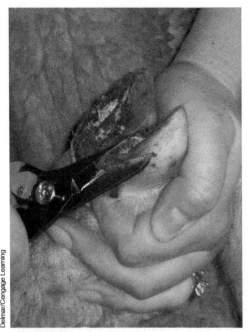

Delmar/Cengage Learning

FIGURE 11-21

Hoof trimming is performed with a pair of specialized nippers.

ewe weighs about 8 to 10 pounds. However, fleeces from Down breeds average less than 8 pounds.

If shearing is delayed until after lambing, ewes should be **crutched**. Crutching is the process of shearing the dirty wool around the vulva and udder. Lambs from crutched ewes are born into a cleaner environment and have an easier access to the udder. Ewes should also be crutched before breeding to aid in copulation.

If structurally correct, most adult sheep will show even wearing of the hooves. Those that do not should undergo corrective hoof trimming to avoid lameness. See figure 11-21. Both the inside and outside hoof walls should be trimmed even with the fleshy center of the hoof. The hooves of rams should routinely be trimmed before the breeding season.

Ideally, each ewe should deliver and wean two lambs. Unfortunately, some ewes lamb only singles while others have triplets. Sometimes, one lamb in a set of twins is **stillborn** (born dead). In such instances, one lamb from a set of triplets may be placed with the ewe that has only one lamb. The adoptive ewe may be convinced to accept the new lamb through several methods. One method involves rubbing the lamb with placental fluids from the adoptive ewe. These fluids hide the odor of the lamb and make the ewe think it is her own. Another method involves skinning a ewe's stillborn lamb and tying the pelt around the new lamb. Again, the odor of her own lamb may fool the ewe into accepting the new lamb.

SHEEP PARASITES

Sheep are subject to a variety of parasites and diseases. Those of economic importance are surprisingly few considering the adage, "a sick sheep is a dead sheep."

WORMS

Parasite infestations in the form of worms are common, especially in sheep on permanent pastures. Often, the sheep ingest worm eggs or larvae while eating pasture grasses. Worms rob sheep of feed nutrients. A heavy worm load causes sheep to lose body

Delmar/Cengage Learning

FIGURE 11-22

Drenching involves placing medication at the back of a sheep's throat with specialized equipment.

condition—even when on a high plane of nutrition. A light worm load can cause a decrease in gain and efficiency. Control of worms is a two-part process. The first is to rotate pastures to break the life cycle of worms. The second is to routinely **drench** or otherwise treat all sheep with an approved wormer. See figure 11-22. Drenching involves placing worming fluid in the back of the throat so the sheep is forced to swallow it. Lambs to be placed on finishing diets should be drenched at weaning.

COCCIDIOSIS

A second parasite that often causes losses in feedlot lambs is coccidiosis. Coccidia also infest cattle, swine, and poultry. Coccidia are microscopic organisms that invade the intestinal lining, reduce nutrient uptake, and cause diarrhea. Coccidia reside in spores when outside the host animal. Spores are resistant to heat as well as disinfectants, and are very difficult to eradicate. Prevention of coccidiosis involves reducing the stress on weaned lambs moved to the feedlot. Feed additives called coccidiostats may be fed in the grain mix during the finishing period to control coccidiosis.

SHEEP DISEASES

CLOSTRIDIA

Clostridia also reside in spores and cause a variety of diseases in sheep. *Cl. chauvoei* causes a disease known as blackleg. Blackleg is relatively uncommon but results in the swelling of various parts of the body. These parts are often the site of a recent wound. *Cl. tetani*, the causative agent in tetanus, also enters through open wounds, such as those associated with castration or shearing. *Cl. perfringens* has two types. Type C causes bloody scours or hemorrhagic enterotoxemia. Bloody scours most often affect young, rapidly growing nursing lambs. Type D causes overeating disease or enterotoxemia. Overeating is so named because it usually affects the biggest, fastest gaining finishing lambs. Animals infected with *Cl. perfringens* types C and D have high mortality rates.

Prevention of diseases caused by *Clostridia* involves vaccination. Vaccines for all Clostridial diseases are packaged separately or in combination. Ewes and lambs should be routinely vaccinated for *Cl. perfringens* types C and D. *Cl. tetani* vaccinations are also advised. Initial immunization requires two doses. Breeding ewes and rams require an annual booster.

PNEUMONIA

Pneumonia is perhaps the most costly disease affecting feedlot lambs. Alternating weather conditions, poor ventilation, dampness, chilling, and poor nutrition contribute to pneumonia. Most pneumonia is caused by a bacterial invasion of the lungs. Symptoms include reduced feed intake, depression, and nasal discharge. Prevention involves reducing the factors that predispose lambs to pneumonia. Attention to comfort will prevent most pneumonia. Treatment with antibiotics is effective for bacterial pneumonia if the disease is detected early.

FOOT ROT

Foot rot is a bacterial infection of the hoof. Foot rot is most troublesome in wet pastures. Lameness is the most common sign, while a distinct rotting odor is also obvious. If foot rot is diagnosed in a flock, all the affected animals should be isolated in a dry environment and infected parts of the hoof should be trimmed. The entire flock should be forced to move through a foot bath containing a proscribed mixture of antibacterial ingredients. Trimming and footbaths should be continued weekly until signs are gone. To prevent reinfection, footbaths are recommended for several weeks after signs have disappeared. Treated sheep should not be restocked onto infected pastures for at least 25 days.

SORE MOUTH

Sore mouth is a disease caused by a viral infection of the lips, tongue, and gums. The infection itself does not cause serious symptoms, but the resulting sores drastically decrease feed intake. Treatment is difficult. Sores normally heal in one to two weeks. Lambs eat small amounts of a soft, palatable feed while sores heal. Vaccination is possible for flocks that have a history of sore mouth. However, it is not recommended for those that have never contracted the disease. Sore mouth is contagious to humans, so managers should wear rubber gloves when treating infected animals.

SHEEP HOUSING

Mature sheep have no special housing requirements. Winter wool allows them to stay warm even in the most inclement weather if a suitable windbreak exists. Many producers provide an open-fronted shed. Recently shorn sheep should be provided a warmer environment in cold, wet weather to help prevent pneumonia.

When lambing season falls during the winter or early spring, ewes are often confined to a lambing shed. Most lambing sheds are enclosed, and some are heated. These sheds allow lambs to be born into a warm, low-stress environment. Ewes are often penned alone during lambing for close observation. This also allows ewes and lambs to develop a firm bond. After several days inside, ewes and lambs are returned to a paddock with access to an open-fronted shelter. In especially inclement weather, lambs can be given access to a more confined shelter with openings too small for ewes.

Finishing facilities are normally a dry lot with access to a shed or cover. Some lambs are finished in confinement on slatted floors built over a deep pit. In such confined situations, mechanical or natural ventilation must be utilized to ensure acceptable air quality.

Sheep are more difficult to fence than cattle. Total fence height should be about 42 inches. Woven, barbed, or mesh wire and board fences are all used with success. The bottom strands of the fence are the most critical as they will be challenged most often. The first strand should be about 5 inches from the ground. The second strand should be about 13 inches from the ground. Electric fences should be charged by a powerful charger, because wool is a poor conductor of

Animal Science Facts

The U.S. military consumed over 4 million pounds of domestic wool in 2008.

Delmar/Cengage Learning

FIGURE 11-23

High-tensile fence serves to keep sheep in and predators out of a pasture.

electricity. Sheep with long fleeces may push through poorly charged electric fences without getting shocked. High-tensile strand fence, five to seven strands (two to four electrified), works well but may be too expensive in many production situations. See figure 11-23.

MARKETING

Feeder lambs can be marketed through local sale barns, tele-auctions, organized feeder lamb sales, or directly to the feedlot. As mentioned previously, eastern producers market many feeder lambs as hothouse lambs. Some of these lambs move through sale barns or country buyers.

Finished lambs are sold through many of the same channels as feeder lambs. Lambs from large feedlots may be marketed directly to packing plants. Some lambs are sold on a weight and grade carcass basis. In such arrangements, adjustments may be made to the price depending on carcass quality. Lamb processing plants are most abundant in Colorado and California. Current prices for lambs can be found in local trade publications or on the Internet.

FIGURE 11-24

Wool value is based on fiber fineness, among other factors.

Wool is marketed through its own distinct channels. Depending on the part of the country and the amount to be sold, wool may be purchased by local or regional dealers, or wool cooperatives. Large amounts of high-quality wool may be sold directly to a woolen mill. The wool's value is based on the fiber fineness, density (crimp), length, and cleanliness. See figure 11-24. The impurities found in wool are called grease.

Wool is graded according to the fineness of its fiber. Three different grading systems are used: blood system, count system, and micron system. The blood system computes a grade based on the percentage of Merino blood (ancestry) in a sheep. The blood system grades are fine, 1/2 blood, 3/8 blood, 1/4 blood, low 1/4 blood, common, and braid. The count system correlates grades based on the number of hanks (560 yards) spun from one pound of wool. Grades range from finer than 80 to coarser than 36. The micron system measures the average fiber diameter. Micron grades range from 17.00 to 40.21 plus. Wool is further divided by use. Fine wools (apparel wools) are used for clothing. Coarser wools (carpet wools) are made into rugs.

The market price of wool has traditionally been supported by the government. The government made up the difference between the actual market price and a predetermined base price. However, the wool price support program ceased in 1996 but resumed with the 2002 Farm Bill and will continue through at least 2012.

INDUSTRY ORGANIZATION

Western range flocks account for about 85 percent of the sheep inventory of the United States. Generally, range flocks are characterized by at least 1,000 head. In range flocks, sheep are most often the primary enterprise.

Outside range country, smaller farm flocks usually maintain fewer than 200 head. Sheep tend to be secondary to other farm enterprises in these operations. Therefore, management is not as critical to overall farm success. Profitable hothouse lambs, fed indoors

FIGURE 11-25

Lambs for 4-H and FFA projects are one niche market in the sheep industry.

over the winter and marketed at Easter and other holidays, have become a **niche market** (specialized) in the sheep industry. Other niche markets include American lamb for high-end restaurants, sale of club lambs for 4-H and FFA projects, and purebred sales to other small purebred producers. See figure 11-25.

INDUSTRY ISSUES

General decline threatens the sheep industry. Four main factors have contributed to this fall. Declines in inventory and the consumption of lamb and mutton have caused great industry concern. In addition, the wide availability of synthetic clothing materials has provided wool formidable competition in fabric selection. As a result, American wool use fell from an average of more than 5 pounds annually during the World War II era, to a current rate of 1 pound or less. However, due to an increased world demand for wool, exports from the United States have increased. Importation of lamb has increased greatly since the 1950s. Much of the lamb found in supermarkets is imported from Australia or New Zealand.

On July 1, 2002, the Lamb Promotion, Research and Information Board began collecting check-off funds at the rate of 1/2 cent per pound of live weight. The estimated $3 million generated each year funds research, promotion, and provide information to promote and expand markets for American lamb. Hopefully the check-off program will be able to reverse the downward trend of the American sheep industry.

Since the 1950s scientists have known that some sheep harbor a disease known as scrapie, which is similar to the illness that causes mad cow disease. In 2002, USDA initiated a national sheep identification program to trace confirmed scrapie cases back to the flock of origin. The goal is to eradicate scrapie from America's sheep herd by 2010. All sheep must carry an official USDA scrapie tag that allows the animal to be traced to its source flock. See figure 11-26.

Another industry issue is predation. Flock profits are reduced by **predators** (animals killing the sheep). In 1994, (the last year for which data are available) over 0.5 million sheep and lambs were lost to predators, amounting to a loss of $17.7 million. Coyotes accounted for over 60 percent of sheep predation.

Finally, the sheep industry, like all other livestock industries, is struggling to comply with the USDA's National Animal Identification System (NAIS) program. The program was put in place to individually identify

Photo courtesy USDA

FIGURE 11-26

All sheep are required to carry an official USDA scrapie ear tag.

Animal Science Facts

Predators such as coyotes and feral dogs have always been a problem for sheep producers. Many western sheep producers graze an alpaca or a llama with each flock of sheep. Alpacas and llamas are very hostile to canines and do an admirable job of protecting their charges.

FIGURE 11-27

An Extension educator teaches market lamb management.

animals and help track livestock to their source in the event of a disease outbreak. Because of the scrapie tag system, the sheep industry may be one of the progressive industries in implementing the NAIS.

CAREER OPPORTUNITIES

As in the other meat animal industries, career opportunities are available in sheep production. However, these jobs are available on a far more limited basis, as lamb and mutton account for less than 1 percent of all meat consumed in the United States. Sheep ranchers often employ shepherds who live with the flocks in range grazing areas for months at a time. Other people shear sheep for a living. Skilled shearers can remove an entire fleece in a few minutes. Careers as **jockeys** (people who buy and sell sheep or other livestock) can also be found. These experts must be skilled at selection and economics if they are to turn a profit for themselves and their clientele.

Additional positions are sometimes offered by cooperatives, which collectively market industry products for better producer profits. Other opportunities are available in lamb and wool grading and processing. These individuals help assure quality products for consumers. As with other livestock commodities, support personnel, such as extension agents, educate producers in newly developed industry technology. See figure 11-27.

SUPPORTING ORGANIZATIONS

The American Sheep Industry Association (ASI) is located in Centennial, Colorado. This federated association, with more than 40 state affiliates as well as individual members, provides a unified voice for the sheep industry. Moreover, the association serves a host of purposes including research, education, promotion, and advocacy. A producer board, consisting of representatives from state affiliates, advises and oversees program direction. The ASI established the National

Sheep Improvement Center in 1996 and played a key role in developing the lamb check-off program.

Two other national organizations support the sheep industry. The National Lamb Feeders Association sponsors educational programs for producers. In addition, the American Sheep Industry Women cooperate with the other groups in advocacy work.

SUMMARY

Sheep production centers in the western range states. Of the nearly 50 sheep breeds in the United States, only nine have significant registrations. These nine breeds are divided among those having different uses in crossbreeding schemes. Breeding systems vary by geographical area. Western range flocks are managed for one lamb crop and one wool **clip** (fleeces from the flock) per year. Some eastern farm flocks use prolific ewes that breed year-round to produce a lamb crop every eight months. Ewes are fed a high-forage diet. The highest nutrient requirements occur during late gestation and lactation, when some grain feeding may be necessary. Some just-weaned lambs are sold directly for slaughter, while others are fed grain for several months before slaughter. Tail docking, yearly shearing, and crutching are examples of management practices used on sheep. Worms, coccidiosis, *Clostridial* infection, pneumonia, foot rot, and sore mouth are common health problems. Facilities for sheep normally consist of a windbreak or open-fronted shed. Totally enclosed lambing facilities are often used. Lambs are marketed through channels similar to other livestock species. Wool is sold by grade.

The sheep industry can be organized into large-range flocks, usually numbering over 1,000 head, and much smaller farm flocks, generally less than 200 head. General industry decline is due to a drop in wool use, reduced consumption of lamb and mutton, development of synthetic fabrics, and importation of wool and lamb. Although careers in the sheep industry exist, they are not as plentiful as the other livestock species. The American Sheep Industry Association serves as the voice for the sheep industry.

CHAPTER REVIEW

EXPERIENTIAL LEARNING OPPORTUNITIES

1. Spend a day with an extension family living educator. Ask if he or she could focus on lamb recipes and discuss the variety of wool fabrics and associated uses.

2. Find out if there is a wool cooperative in your area. If so, volunteer to assist the day wool is pooled for grading and sale.

3. Assist a wool judge at a county fair.

DEFINE ANY TEN KEY TERMS

clip

creep feeding

crutched

docked

drench

fine-wool breed

flushing

hothouse lamb

jockeys

medium-wool breed

niche market

predators

scur

seasonal breeder

stillborn

synthetic breed

QUESTIONS AND PROBLEMS FOR DISCUSSION

1. The Down breeds were developed in Southern _____.

2. Classify the following breeds as a medium or fine wool:

   Rambouillet—

   Hampshire—

   Suffolk—

3. List the two synthetic breeds discussed in this chapter.

4. Name three breeds often utilized as range ewes.

5. The major source of income for the Eastern and Midwest sheep producers comes from the sale of _____.

6. Write the formula for calculating lambing percentage.

7. Explain when nutritional demands on ewes are the greatest.

8. Lambs consuming a finishing ration can be expected to gain _____ pound(s) each day.

9. Fat cover on slaughter lambs should range between _____ and _____ inch(es).

10. What are two benefits of crutching ewes?

11. Discuss three methods of pneumonia prevention.

12. What pasture condition aggravates foot rot?

13. True or False. Sheep with long fleeces may be able to push through an electric fence without getting shocked.

14. Name three factors used to determine wool value.

15. The government supported wool prices until _____, but reinstated wool payments in the 2002 Farm Bill. (List the year)

16. In 2007, sheep inventory in the United States numbered _____.

17. Consumption of lamb and mutton in the United States averages less than _____ pound(s) per year.

18. _____ ranks first among states in receipts generated by sheep and lambs.

19. In 2007, _____ million pounds of wool were produced in the United States.

20. Range flocks usually number more than _____ head.

21. List factors that have contributed to the sheep industry decline.

22. Predators cost the sheep industry _____ dollars in 1994.

23. Lamb and mutton account for less than _____ percent of the meat consumed in the United States.

24. What does a jockey do?

25. Where is The American Sheep Industry Association located?

26. True or False. The American Sheep Industry Association is affiliated with 20 state-organized groups.

27. The lamb check-off is set at a rate of $0.005 per _____ of live weight.

28. _____ graders evaluate the fineness, crimp, and cleanliness of the sheep product.

29. Sheep producers can benefit from educationally oriented programs sponsored by this organization. Name the organization.

30. What is the primary function of the American Sheep Industry Women?

12

Horse Management and the Horse Industry

HI HO SILVER—AWAY!

Objectives

After completing this chapter, students should be able to:

▶ Give an overview of the horse industry in the United States

▶ Categorize the characteristics of the major breeds of horses

▶ Compare the biology of horse breeding with that of other livestock species

▶ Develop a feeding program for a pleasure horse

▶ List four management practices used in the horse industry

▶ Describe common diseases that affect horses

▶ Make a chart of preventive health care measures used in horses

▶ Discuss the housing requirements for horses

▶ Explain current issues in the horse industry

▶ List three careers in the horse industry

▶ Name two organizations that play supporting roles in the horse industry

Key Terms

AHC	draft	founder
anthelmintic	dressage	hand mating
band	encephalomyelitis	longeing
colic	equestrian	sulky
distemper	farrier	tack

Career Focus

Ben grew up in Amish country and appreciated the look and sound of a good carriage horse as it clip-clopped down the road in front of his house. One day while on his way back from high school, Ben spied a man at the neighboring Amish farm bending over the front leg of a horse, holding its foot in the air. Ben went to investigate and met Jim, the local **farrier** (hoof caretaker). Ben watched as Jim carefully picked up the big horse's foot, filed the hoof, and installed a brand new horse shoe, the source of the sound Ben enjoyed so much. Jim described the details of the job to Ben who became more and more interested in the work. When Jim was finished, Ben asked if he could ride along and help shoe horses with Jim the next Saturday. Ben went on to become an apprentice farrier and eventually took over Jim's business of shoeing horses for a living.

INTRODUCTION

Horses and humans have a long history together. As one of the first domesticated animals, horses have provided transportation, draft power, companionship, and even meat for their human caretakers. Until the advent of tractors and automobiles, nearly everyone used horses and knew how to care for them. Today, horses are used almost exclusively for recreational purposes, including racing, riding, and exhibition. Although horses are not as necessary to society now, their popularity remains strong. Today most horses in the United States are used for sport or pleasure, although some workhorses remain active on ranches and in Plain Societies, such as the Amish. The horse industry does not organize as easily as swine, cattle, or sheep. Moreover, interests and career opportunities in the horse industry vary greatly. Unwanted horses, shortage of veterinarians, and animal identification issues concern the horse industry.

FIGURE 12-1

In this chapter you'll learn why a horse's mouth is a good place to look.

© Studio 37. Used under license from Shutterstock.com

HORSE INDUSTRY OVERVIEW

According to the American Horse Council (**AHC**), over 80 percent of all horses in the United States are pleasure horses, encapsulated by horse racing,

FIGURE 12-2

Rodeo horses combine with police and work horses to represent 20% of the American horse population.

showing, and recreation. The other 20 percent represent work, rodeo, and police horses, and other uses. See figure 12-2. In 2005, the AHC estimated that 9.2 million horses in the United States generated over $39 billion of goods and services annually. Horses are raised throughout the country with heavy populations of breeding stock located in the Kentucky Bluegrass region.

BREEDS

Individual breeds of horses are genetically the most closely related of the livestock species. Most breeds descend from, or are composites of, a handful of original breeds. To compound matters, some breeds are classified by color. Any horse of a certain color or color pattern can be registered as a member of one of the colored breeds, though its genetics may be of an entirely different breed.

Horses are loosely divided by size into three main categories: light horses, heavy (**draft** or work) horses, and ponies. Light horses are further sorted by use. Some breeds are classified as riding horses, which may be used for pleasure riding, as cattle horses, or as mounts for equine sporting events. Other breeds

are categorized as racehorses, either mounted or harnessed to pull a **sulky** or cart. Still other breeds are classified as driving horses, distinctly developed for pulling carriages.

LIGHT HORSE BREEDS

Although many light horse breeds exist, a brief discussion of some popular breeds follows.

Quarter Horse

The Quarter Horse is the most popular breed in the United States. Nearly 116,000 Quarter Horses were registered in 2008. See figure 12-3. Quarter Horses are heavily muscled animals that excel in sprinting short distances. The name, Quarter Horse, originates from accomplishments in quarter mile races. The development of the Quarter Horse breed relied heavily on Thoroughbred bloodlines. However, the breed was entirely developed in the United States. The American Quarter Horse Association (AQHA) was first formed in 1940. Quarter Horse coat color often ranges in shades of brown or red. Palominos (yellow or golden color with light

Photo courtesy of the American Quarter Horse Journal

FIGURE 12-3

Quarter Horses are the most popular breed in the United States. Their name is derived from their accomplishments in quarter-mile races.

Animal Science Facts

A Horse of a Different Color

Horse colors such as black, white, and gray are self-explanatory. However, some other color names are a bit more vague.

Albino = white

Appaloosa = mottled over rump and loin

Bay = brown with black points (legs, mane, and tail)

Blue Roan = dark and white hairs intermixed

Buckskin = tan with black points and dorsal stripe

Chestnut = deeper reddish brown than sorrel

Sorrel = reddish brown

Paint = distinct patches of dark and white coloring

Palomino = golden

Red dun = reddish tint with dark points

Red or Strawberry Roan = red and white hairs intermixed

colored manes), blacks, and roans are also common. Quarter Horses are usually ridden as cattle horses or pleasure horses.

Thoroughbred

Thoroughbreds are among the most popular breeds with registrations totaling 36,600 in 2008. See figure 12-4. Known for their speed over long distances, Thoroughbreds were developed in England in the 1600s. The first Thoroughbred was imported to America in 1730. One of the oldest breeds of horses, Thoroughbreds have been used as foundation stock in the formation of many other breeds, both in the United States and abroad. It is estimated that over 75 percent of all horses carry some Thoroughbred blood.

Thoroughbreds vary in color from many shades of brown, to black and even gray. They are refined, angular animals with long legs. Thoroughbreds are built for speed and used as racing horses.

FIGURE 12-4

Due to their angular refinement, Thoroughbreds are without peer as racehorses.

Arabian

As its name implies, the Arabian breed was first developed in the desert region of Arabia. See figure 12-5. The natives of the area were warring nomads who bred horses to quickly travel over long stretches of open country. These horses were prized for their alert attitude and athletic endurance. Smaller than the Thoroughbred, Arabians are primarily used as a riding horse. Arabians were first imported to the United States in 1765. The Arabian Horse Registry was established in 1908. In 2008, 6,120 Arabians were registered in the United States.

Paint Horse

Paint Horses, distinguished by their splotched coat color, classify as a colored breed. See figure 12-6. White areas on the hide of the Paints are interspersed with large areas of another color. The colored areas are much larger in comparison to the color patterns of the Appaloosa (discussed later in this chapter). Paints trace their ancestry to nondescript spotted horses found in America. To meet registration requirements, Paints must be descended from registered Paints, Quarter Horses, or Thoroughbreds. Paints are

FIGURE 12-5

Arabians are generally recognized as the oldest horse breed.

FIGURE 12-6

Famed in folklore, Paint Horses were commonly used by Native Americans.

commonly used as riding horses, cattle horses, or pleasure horses. The American Paint Horse Association was formed in 1965. There were 29,534 Paint Horses registered in 2008.

Standardbred

Standardbreds were developed in the United States in the early 1800s through Thoroughbred crosses. See figure 12-7. They were originally used as driving horses in Colonial America. Today, Standardbreds often compete as harness racehorses. The Standardbred earned its name for the ability to move in two gaits, the trot and the pace, at the same or standard speed. Standardbreds come in the same color patterns as Thoroughbreds but are smaller and more muscular. There were 9,510 Standardbreds in the registry as of 2008.

Photo courtesy of the United States Trotting Association

FIGURE 12-7

Standardbreds were developed primarily for speed at the trot and pace.

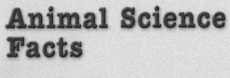

Animal Science Facts

Horse Gaits

Walk = slow, four-beat gait

Trot or jog = rapid, two-beat gait (opposing front and rear hooves move in unison)

Pace = rapid, two-beat gait (front and rear hooves on the same side move in unison)

Canter or lope = slow, three-beat gait

Stepping pace = similar to the pace but hooves on the same side do not strike the ground at exactly the same time. Performed by five-gaited horses

Running walk = smooth, four-beat gait performed by five-gaited horses

Rack = quick, showy, four-beat gait performed by five-gaited horses

Fox trot = choppy trot

Appaloosa

The ancestors of the modern Appaloosa first came to Mexico with Spanish explorers in the 1600s. See figure 12-8. Later, Native Americans in the Pacific Northwest acquired this spotted strain of horse. Appaloosas are mottled dark and white, have black and white striped hooves, and have light pigment around the eyes. The Appaloosa Horse Club was formed in 1938 to preserve the remnants of these beautiful horses. Today, most Appaloosas are ridden for pleasure. There were 5,390 Appaloosas registered in 2008.

Tennessee Walking Horse

The Tennessee Walking Horse or Tennessee Walker was developed in the Tennessee Valley in the 1800s. See figure 12-9. Based primarily on Thoroughbred blood from a variety of sources, the Tennessee Walker executes a gait called the running walk. Walkers carry this unique gait smoothly,

Photo courtesy of the Appaloosa Horse Club Inc.

FIGURE 12-8
Appaloosas were first brought to North America by Spanish explorers in the 1600s.

Photo courtesy of the Tennessee Walking Horse Breeders Association ©Vesty

FIGURE 12-9
The Tennessee Walking Horse is a rugged breed with outstanding stamina.

easing the physical stress on the rider. Currently, Tennessee Walkers are exhibited as riding horses and are ridden for pleasure. The breed association was first formed in 1935, and 6,820 horses were recorded in 2008.

Morgan

Morgan horses were the first truly American breed. See figure 12-10. They were developed in the late 1700s from a single stallion called Justin Morgan. Justin Morgan, named after his owner, is thought to have been of Thoroughbred and Arabian breeding. Morgans, known for their versatility, were an asset on the Eastern frontier. Their short, round appearance originally lent itself to work as a draft animal. However, Morgans were also used as riding horses and were even raced occasionally. Morgans have been useful in the development of subsequent American horse breeds, such as the Standardbred and Tennessee Walker. There were 2,512 Morgans registered in 2008.

Pinto

Pinto Horses are a true color breed. See figure 12-11. Despite size or breed, almost any horse or pony with a spotted color pattern can be registered as a pinto. The pinto color pattern originated from horses of Spanish extraction. Most Pintos are currently used for pleasure or as cattle horses. There were 2,973 Pinto Horses registered in 2008.

Photo courtesy of Bentley Photography

FIGURE 12-10

The first truly American breed, Morgan horses were developed from a single outstanding stallion named Justin Morgan.

FIGURE 12-11

The Pinto color pattern originated from horses of Spanish extraction.

American Saddlebred

American Saddlebreds were developed in the eastern United States in the mid-1800s. See figure 12-12. They are used as a smooth gaited saddle horse. Many American Saddlebreds are five-gaited horses (see animal science fact on horse gaits). The background of American Saddlebreds consists of mainly Thoroughbreds as displayed by their large size. Today, American Saddlebreds are used primarily for pleasure riding or show. The American Saddlebred Horse Association was formed in 1891 and registered 2,811 horses in 2008.

Palomino

Palomino horses descended from golden-colored horses brought to the United States by early Spanish explorers. See figure 12-13. The mating of two Palominos does not necessarily result in a Palomino offspring. For instance, the mating of two Palominos will result in only 50 percent Palomino offspring. However, mating Chestnut-colored horses to albino horses results in 100 percent Palomino colored offspring. Only horses of Palomino color can be registered as a Palomino. Two breed associations, established in the 1930s and 1940s, exist. Palominos are currently used for pleasure riding and cattle purposes. The Palomino Horse Breeders of America registered 1,055 horses in 2008.

Photo courtesy of the American Saddlebred Horse Association

FIGURE 12-12

Although American Saddlebreds are extremely alert and curious, they are highly intelligent and people oriented.

Photo courtesy of the American Quarter Horse Association

FIGURE 12-13

Palominos are golden or yellow in color with light-colored manes.

DRAFT HORSE BREEDS

The preceding breeds are all recognized as light horse breeds, mainly used for transportation. The following are heavy or draft horse breeds developed for use as work animals. All draft breeds descended from the native Flemish horse, a low set, heavily muscled animal. Draft horses peaked in popularity before the advent of farm tractors. In those days, draft horses pulled plows and other agricultural equipment. Draft horses are still used today by the Amish, much the same as in the early 1900s.

Belgian

The Belgian breed was developed in Belgium and was first introduced into the United States in 1886. See figure 12-14. Belgians are the most massive of the draft breeds. Mature stallions can weigh over a ton. Most Belgians are chestnut, roan, or bay in color. The Belgian Draft Horse Corporation of America was established in 1887 and had registered a total of 2,508 registered horses as of 2008.

Percheron

Originating in France, the Percheron was first brought to the United States in the 1840s and 1850s. See figure 12-15. Percherons are smaller than Belgians

Photo courtesy of USDA

FIGURE 12-14

Belgians are by far the most numerous of all draft horse breeds in the United States.

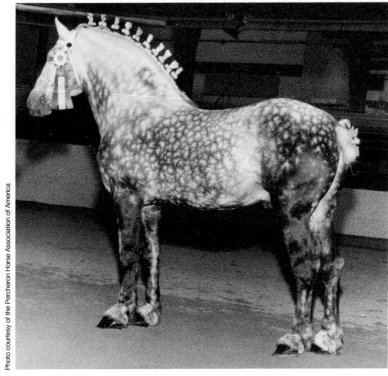

Photo courtesy of the Percheron Horse Association of America

FIGURE 12-15

Percherons were the first draft breed to be imported. For their size, they are remarkably agile.

and are predominantly black or gray. Registrations totaled 1,760 in 2008.

Clydesdale

Clydesdales, originating in Scotland, are perhaps the best-known draft breed to the American public. See figure 12-16. Clydesdales were imported into this country in the 1870s. They are not as heavy as the Belgian and Percheron. Most are bay and brown with white markings, but blacks, grays, chestnuts, and roans are occasionally seen. Registrations totaled 536 in 2008.

PONIES AND MINIATURE HORSES

According to recent registrations, several breeds of ponies show limited popularity. Ponies are less than 58 inches (14.2 hands) tall at the withers. See figure 12-17. National data for pony registration numbers are not available. Miniature Horses, which must meet stringent size requirements for registration (less than 30 inches at the withers), are exhibition

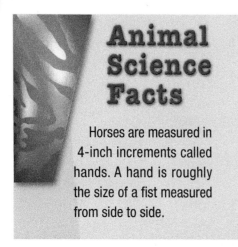

Animal Science Facts

Horses are measured in 4-inch increments called hands. A hand is roughly the size of a fist measured from side to side.

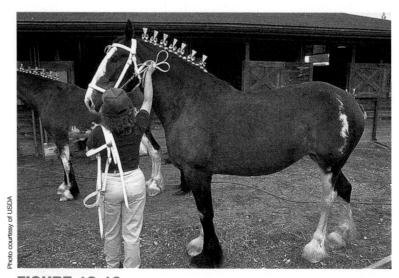

Photo courtesy of USDA

FIGURE 12-16
A Scottish breed, Clydesdales are noted for straight and high action.

Photo courtesy of the American Shetland Pony Club

FIGURE 12-17
Shetlands are among the most popular pony breeds.

animals popular in petting zoos and parks. There were 7,496 Miniature Horses registered in 2008. Ponies and Miniature Horses differ in their body proportions.

(Registration numbers provided by the National Breed Associations and the American Horse Council.)

 BREEDING

Large-scale horse breeding is a highly specialized field. There are few producers in relation to other species of livestock. Farms and ranches in the business of breeding horses must have a genetically sound breeding herd and a planned marketing strategy for selling progeny.

Some herds may be in the business of producing registered Thoroughbred racehorses to train or to sell to investors. Other breeders may breed a specific sire to a **band** (herd) of broodmares. See figure 12-18. That sire may be selected for his proven abilities in **equestrian** (horse and rider) events or harness races. Progeny from this type of breeding operation are not necessarily registered or purebred horses but are simply bred to perform a desired task.

The biology of horse breeding is similar to other species of livestock. Horses are seasonal breeders like sheep. An increase in day length triggers the onset of estrus in mares. If left to natural means, heat periods begin in March or April and cease in September or October. Mares normally cycle every 21 to 23 days during the breeding season, but some mares are very irregular. Many producers use artificial lighting to trick mares into coming into heat at other times of the year. Heat detection for mares was described in Chapter 5.

The mating of a receptive mare is normally done naturally, either in a pasture setting or, more often, through **hand mating** (supervised mating). During hand mating, a handler prepares the mare for breeding by washing her vulva and wrapping the tail to keep it out of the way. The mare's rear legs are often hobbled during breeding so the mare does not move or kick and risk injury to the stallion. The mare is tied

FIGURE 12-18

Band of broodmares in the Western plains.

Photo courtesy of USDA

FIGURE 12-19

Horses are seasonal breeders with most foals dropped in April, May, and June.

to a stall and the stallion is brought to the mare. The handler normally assists in copulation.

Artificial insemination (AI) technology is available for horse breeders. However, some breed associations will not allow the registration of progeny produced by AI. Other associations will allow registration, but semen may not be frozen or transported from where it was collected. AI greatly reduces the chance of injury to both the stallion and mare during mating. It also allows more mares to be mated to a single, high-quality stallion.

Gestation period for horses is about 11 months. Most foals are dropped in April, May, and June. See figure 12-19. Mares exhibit a foal heat one to two weeks after foaling. Most producers wait until the second heat, 25 to 30 days after foaling, to rebreed mares. This period allows the mare time to recover from the stress of foaling.

HORSE NUTRITION

Survey 100 people on the street about what horses eat and the most likely answer will be oats. The second most popular answer will probably be hay. These two are excellent horse feeds, but others are routinely fed to horses as well.

As mentioned in Chapter 4, horses are post-gastric fermenters. The first part of the digestive tract is similar to that of a human or pig. However, the cecum of the horse is developed for the digestion of forages. Thus, unfermented forages are a preferred feedstuff for horses.

Good quality pasture is preferred as a forage for horses. See figure 12-20. Bluegrass, orchardgrass, bromegrass, fescue, and reed canarygrass function as good pasture species. Legumes, such as alfalfa, trefoil, or clover, may be used alone as pasture or mixed with grasses. When stored forage must be fed, hay made from any of the above pasture species makes excellent feed if cut at a young stage of maturity. Many equestrians prefer young timothy hay, but timothy makes poor pasture.

Grains often added to horse diets include oats, barley, corn, wheat, and milo. Protein supplements include linseed meal, cottonseed meal, canola meal, and, of course, soybean meal. Bran is often used as a source of fiber to maintain digestive tract function.

Nutrition for breeding mares, foals, and yearling horses is similar to other species of livestock. Mares in late gestation and lactation have the highest nutrient demands and require top-notch nutrition. Both quantity and quality of feed are critical. Bran is often used in mare diets around foaling time to prevent constipation.

© Geka. Used under license from Shutterstock.com

FIGURE 12-20

Good quality grass pasture makes fine feed for horses.

Foals should be creep fed beginning at two weeks of age. The nutrient density of mares' milk decreases drastically one to two months after foaling.

After weaning, foals must continue to grow and develop at a rapid pace. A 400-pound, weaned foal should consume about 8 pounds of hay and 6 pounds of grain per day. Growth requirements indicate the need for high levels of protein and good quality hay or pasture. Rations should be gradually increased as the foal grows through its first winter.

Developing yearlings can survive on good quality pasture in their first summer, but many producers supplement pasture with a small amount of grain. Yearlings entering their second winter will require about 8 pounds of grain and 12 pounds of hay per day. See figure 12-21.

Feeding pleasure horses presents some challenges. The nutrient needs of mature pleasure horses depend on the animal's size and level of activity. Work is divided into three categories. Light work is less than three hours of riding per day. Horses worked lightly need only ½ pound of grain and 1½ pounds of hay per day per 100 pounds of body weight. Medium work is three to five hours per day. Medium-worked horses require 1 pound of grain and 1 pound of hay per 100 pounds of body weight. Heavy work is more than five hours per day. Horses worked heavily require 1½ pounds of grain

Photo by Jennie Osborne

FIGURE 12-21

Yearling horses require good nutrition to maintain adequate growth.

and 1 pound of hay per 100 pounds of body weight. The total amount of feed and the proportion of grain versus hay always increase with increased demand for energy.

Salt, vitamins, and minerals should be included in the grain mix or provided free-choice for horses of all ages.

Horse diets should be changed slowly and gradually. A gradual adaptation from one grain mix or forage to another over 7 to 10 days reduces the risk of an upset digestive system. Horses should initially be released onto lush spring pasture for short periods. Grass founder, a serious hoof ailment, can the a consequence of overconsumption lush pasture overconsumption.

HORSE MANAGEMENT

Identification of horses was traditionally difficult. If a horse was stolen, the only way an owner could identify it was through some distinguishing physical marking. Some horses were hot branded (hot iron applied to the hide leaving a permanent scar) at weaning as a means of permanent identification. More recently, freeze branding (applying a cold iron to the hide, which causes hair to become white), and lip tattooing have become more popular identification methods.

The old saying, never look a gift horse in the mouth, is sound advice because a horse's teeth can reveal its age. A gift horse may be very old and of little value. However, since it is a gift, it is best to accept it as is. If you are buying a horse, looking in the mouth is very important. Unscrupulous horse traders have made a good living passing off old horses to novices (those new to business). Temporary teeth appear in foals at a very young age. The entire complement of eight temporary teeth is in place by 10 months of age. These temporary teeth wear until about age 2½ years when the front set of permanent teeth replace the front set of temporary teeth. From age 2½ until 4, all eight temporary teeth are replaced by eight permanent teeth. At age 4 or 5, canine teeth appear behind the teeth already present. From that time on, these permanent teeth show varying degrees of wear from which an experienced horse person can determine the age of any horse. See figure 12-22. Also, the angle of the teeth, when viewed from the side, changes with age. Teeth of young horses are nearly

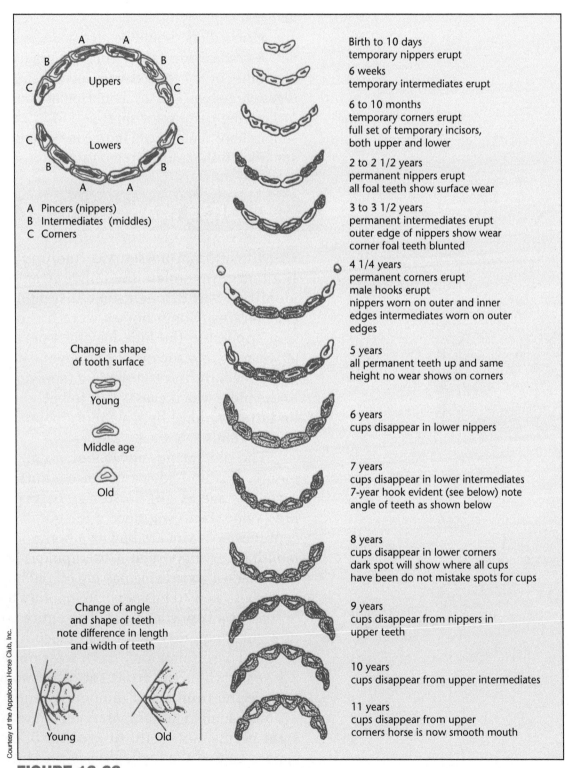

A Pincers (nippers)
B Intermediates (middles)
C Corners

Change in shape of tooth surface

Young

Middle age

Old

Change of angle and shape of teeth note difference in length and width of teeth

Young   Old

Birth to 10 days
temporary nippers erupt

6 weeks
temporary intermediates erupt

6 to 10 months
temporary corners erupt
full set of temporary incisors, both upper and lower

2 to 2 1/2 years
permanent nippers erupt
all foal teeth show surface wear

3 to 3 1/2 years
permanent intermediates erupt
outer edge of nippers show wear
corner foal teeth blunted

4 1/4 years
permanent corners erupt
male hooks erupt
nippers worn on outer and inner edges intermediates worn on outer edges

5 years
all permanent teeth up and same height no wear shows on corners

6 years
cups disappear in lower nippers

7 years
cups disappear in lower intermediates
7-year hook evident (see below) note angle of teeth as shown below

8 years
cups disappear in lower corners
dark spot will show where all cups have been do not mistake spots for cups

9 years
cups disappear from nippers in upper teeth

10 years
cups disappear from upper intermediates

11 years
cups disappear from upper corners horse is now smooth mouth

Courtesy of the Appaloosa Horse Club, Inc.

FIGURE 12-22

A horse's age can be determined fairly precisely by tooth wear.

perpendicular, while teeth of older horses angle forward. Upper and lower teeth should meet evenly. An overbite or underbite is considered an undesirable conformation trait.

Horses on pasture normally get enough exercise. However, horses confined in stalls must have some exercise every day or two to remain healthy. If horses are not ridden or hitched, they may be exercised using a **longeing** rope. Longeing a horse consists of attaching one end of the rope to the horse's halter. A caretaker holds the other end at the center of a circle. The horse is then allowed to exercise around the perimeter of the circle while the caretaker retains control. See figure 12-23.

Foot and hoof health and maintenance keep horses sound and functional. Horses must have healthy feet and hooves to provide locomotion for humans. Hoof trimming should be done every six to eight weeks by an experienced farrier. See figure 12-24. The need for trimming depends on the amount of hoof wear and the presence or absence of shoes. A good farrier can correct damaged or uneven hooves. Horses residing on rocky ground or horses asked to travel on roads should have shoes on their hooves. Shoes protect the hoof from excessive wear and deterioration. Hooves should be inspected and cleaned daily to avoid infections and to spot possible causes of hoof damage.

Photo by Frank Flanders

FIGURE 12-23

Longeing a horse allows it to exercise in a small area.

Photo courtesy of Wayne Nunn Farrier Services. Photo by Frank Flanders

FIGURE 12-24

A good farrier can keep a horse's feet in A-1 condition.

HORSE PARASITES

Parasites in horses are similar to those of other livestock species.

FLIES

Flies cause considerable discomfort to horses. The elimination of fly breeding areas and good sanitation help keep fly numbers at manageable levels. Sometimes horse owners cover a horse's face with a see-through mesh net to keep flies from bothering the horse's eyes.

MANGE

Mange mites infestation can cause serious itching. Mange can be prevented by avoiding contact with infested horses. Mange can be treated with a variety of insecticide powders and sprays.

RINGWORM

Ringworm infests horses as well as cattle. It is most prevalent in winter when horses are confined to stalls. Horses infested with the ringworm fungus show round circular, hairless patches. Individual patches should be treated with an antifungal cream or iodine.

TICKS

Ticks rob horses of nutrients by tapping the horse's blood supply. They can also harbor and transmit other horse diseases. The blood-filled bodies of ticks are easily seen protruding from the horse's hide. Ticks can be controlled by many topical insecticides.

WORMS

Many types of worms infest horses, stealing feed nutrients, which reduces body condition and energy level. Horses should be wormed with an approved **anthelmintic** (worming medicine) every two months to keep worms under control. Worms can build up a resistance to an individual wormer, so anthelmintics containing different active ingredients should be used in rotation. Pastures should also be rotated to break the life cycle of worms.

Large worm loads in the digestive tract can result in **colic**, or abdominal pain. Worms cause colic by essentially plugging up the digestive system. Other causes of colic include a rapid change of diet, too much grain in a single feeding, or too much water after a period of heavy work. Colic can even be caused by improper tooth wear, which prevents the horse from chewing properly. A horse with colic may act nervous, kick at its side, or attempt to roll over. Colic can be prevented by frequent wormings, tooth inspection, and close attention to the horse's diet. Horses with colic should be treated by a veterinarian.

HORSE DISEASES

Horse diseases are somewhat unique to those found in other livestock. Several are transmitted by insects.

ENCEPHALOMYELITIS

Encephalomyelitis (en-sef-a-lo-my-el-itis), also called sleeping sickness, is a horse disease transmitted by mosquitos. See figure 12-25. The disease infects the central nervous system causing the horse to act sluggish and sleepy. Death is common. Encephalomyelitis is easily prevented through an annual vaccination of all horses.

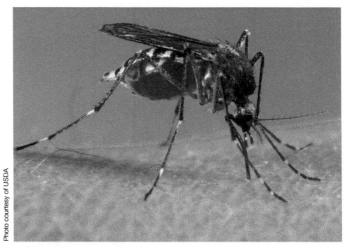

FIGURE 12-25

Mosquitoes spread Encephalomyelitis among horses.

EQUINE INFECTIOUS ANEMIA

Equine infectious anemia (EIA) is a serious viral disease of horses transmitted by insects. Infected horses act depressed, lose weight, and may die. Those that live act as reservoirs for the virus, allowing it to be spread to other horses. The Coggins Test can detect equine infectious anemia before symptoms appear. Horses that test positive are quarantined, permanently identified as a carrier, and often destroyed. Horses sold across state lines must have a recent negative Coggins Test. Yearly testing of all horses is recommended.

EQUINE INFLUENZA

Equine influenza is a viral disease that attacks the respiratory system, causing loss of appetite, depression, cough, and nasal discharge. It is usually associated with an elevated temperature. Equine influenza is common in places where horses from different farms are gathered, such as races and shows. Prevention consists of two initial vaccinations administered one to three months apart followed by an annual booster.

FOUNDER

Founder, or laminitis, is an inflammation of the hoof caused by overconsumption of water, grain, or lush forage. Besides tender feet, a high temperature is usually associated with this ailment. Founder is prevented by restricting feed consumption to the amount required for maintenance plus growth or work. Water should be given a little at a time, especially in hot weather.

POTOMAC HORSE FEVER

Potomac horse fever is a bacterial disease causing high temperatures and severe diarrhea. Death losses can be significant. In areas where the disease is prevalent, a two-dose vaccination given three weeks apart is recommended. Treatment with tetracycline antibiotics is effective if a definite diagnosis has been made.

RABIES

Rabies infections in horses are common but can easily be prevented. Bites from infected animals, such as raccoons, skunks, or dogs, lead to violent behavior or, later, paralysis. Two initial vaccinations administered one month apart and an annual booster help protect against this viral disease.

DISTEMPER (STRANGLES)

Distemper or strangles most often affects young horses and is caused by *Streptococcus* bacteria. The bacteria invade the lymph glands under the jaw. The glands enlarge and may break, producing pus. See figure 12-26.

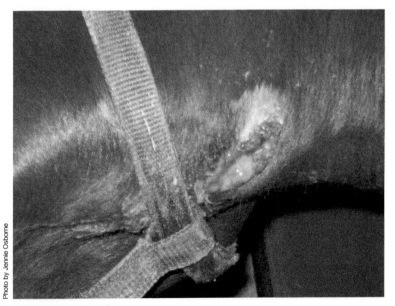

Photo by Jennie Osborne

FIGURE 12-26
Strangles abscesses can rupture.

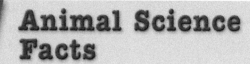

Animal Science Facts

Horse Facts

Average body temperature = 100.5 degrees F

Average age of mares at first mating = two or preferably three years

Gestation period = 11 months

Time to run a quarter mile = 21 seconds

Number of pasture acres per horse = one to five

The disease is often accompanied by a cough and high fever. Distemper can be prevented through immunization every 6 to 12 months.

TETANUS

Tetanus is caused by the bacteria *Clostridium tetani*, which invades the body through open wounds. Initial symptoms include stiffness and then progress to severe muscle spasms. Treatment is difficult, but prevention can be accomplished through two initial vaccinations spaced three weeks apart and accompanied by an annual booster.

WEST NILE VIRUS

West Nile Virus, a mosquito-borne disease, can cause nervous system disorders in infected horses. This virus can also be transmitted to humans through mosquito bites, but horse-to-horse or horse-to-human transmission has not been documented. Horses can be immunized against West Nile Virus, but horse owners should also make attempts to eliminate mosquito breeding habitat around horse boarding areas.

Some horse vaccines are sold in combination, reducing the number of shots required.

HORSE HOUSING

Horse housing generally comes in two types: open shed and box stall. Open sheds are often used for horses having access to pasture or an exercise lot.

The open face of the shed should be to the south or east so that the shed can be used as a windbreak. See figure 12-27. Open sheds should be constructed on high ground so that water from rain or melting snow drains away. The shed should be at least 8 to 9 feet high. About 150 square feet of floor space should be provided per horse. A feed storage area may be built above, or as an addition to the shed.

Box stalls are constructed inside a barn. See figure 12-28. Generally square and 12 feet × 12 feet square in size, the walls of box stalls are made of horizontal planking to about 5 feet high. Above the horizontal planking are vertical bars making the total stall height at least 7 feet. Box stalls normally have

Delmar/Cengage Learning.

FIGURE 12-27
An open horse shelter facing East.

Photo by Jennie Osborne

FIGURE 12-28
This commercial horse facility boasts box stalls for each animal.

clay floors. Stalls should always be clean, dry, and adequately bedded.

Proper ventilation of box stalls should be given serious consideration. A constant supply of fresh air keeps humidity at acceptable levels, reducing the incidence of respiratory diseases. Adequate ventilation also reduces possible disease spread between horses.

In many horse barns, box stalls line both sides of an interior alleyway. In the center of the barn, one or two box stalls may be replaced by a **tack** (harnessing, riding, and grooming equipment) and feed room. Hay is often stored on the second floor.

Horses are most often fenced with board, welded pipe, woven wire, or pole fences. Vinyl-covered high-tensile wire fences, which simulate painted boards, require less maintenance than traditional fencing materials.

MARKETING

A majority of pleasure horses are bought and sold independently and locally through personal contacts and advertisements. However, in the racing industry, high-profile annual sales are conducted where yearling colts and fillies are auctioned to the highest bidder. Yearling horses with impressive racing pedigrees fetch remarkable prices.

INDUSTRY ORGANIZATION

Unlike the meat animal and dairy industries, the horse industry does not neatly organize into type and size of operation. However, horses can be roughly divided into light, draft, and pony categories, based on the size of the horse. Categorizations can also be made on horse use as pleasure horses or working horses.

INDUSTRY ISSUES

In 2006, Congress passed a law banning the slaughter of horses in the United States. Though well intentioned, this law has caused complications for horse

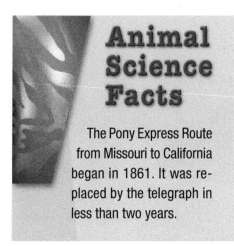

owners. Unwanted horses now have no economic value and horse owners have few options for selling them. In some cases, owners cannot afford to have horses euthanized or to pay a hefty fee for carcass disposal, so they simply release the horses to starve. The AHC has addressed this issue through the Unwanted Horse Coalition whose purpose is to reduce the number of unwanted horses and promote their welfare.

There have been reports of a looming shortage of food animal veterinarians in the United States. Although horses are not food animals, the potential exists for a critical shortage of horse veterinarians in the future caused by the expense of veterinary school and the availability of more lucrative careers for veterinarians treating smaller pets.

Speculation has abounded in the horse industry about the potential for mandatory horse identification. These rumors have initiated fear and misunderstanding among many horse owners. In fact, horse identification under the National Animal Identification System is part of a larger initiative wherein animals are permanently identified to protect livestock and the public from disease outbreaks and agro-terrorism.

CAREER OPPORTUNITIES

Diverse opportunities are available in the horse industry. Horse stables range in size from a few horses boarded for extra income to large-scale operations with a menu of services available to clients. Stables can provide jobs for attendants, barn managers, grooms, trainers, and riding lesson instructors. See figure 12-29. Farriers and veterinarians are usually contracted to care for horses. Sales positions are available to offer feed and tack to horse owners.

At tracks, owners, exercisers, trainers, jockeys, and other track personnel collaborate in the multibillion dollar industry of horse racing. Along with racing, other equestrians may earn incomes by competing with horses. Rodeo and other performance events provide competitors with an opportunity to win prize money. Only the very accomplished can support themselves with such winnings.

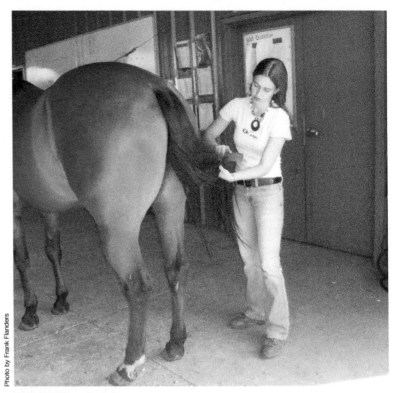

Photo by Frank Flanders

FIGURE 12-29

Many careers in the horse industry begin by simply grooming horses.

Photo courtesy of USDA

FIGURE 12-30

Dressage events are a popular competition for horse and rider.

SUPPORTING ORGANIZATIONS

The American Horse Council represents the horse industry to Congress. However, most national horse organizations are interest centered. The Professional Rodeo Cowboys' Association and the Women's Professional Rodeo Association are just two of several groups serving the rodeo industry. In addition, separate event organizations exist for such rodeo sports as barrel racing and team roping. The United States Equestrian Federation brings together those interested in the equestrian sports. Many breed organizations sponsor their own show committees. Other equestrians interested in certain show aspects such as jumping, pleasure riding, youth (4-H or FFA members), **dressage** (guiding of a horse through unique maneuvers using slight movements), or games form groups to organize show events. See figure 12-30. Other horse-related activities, such as fox hunting and trail riding, have localized organizations supporting these interests.

Horse racing remains one of the most popular sports in the United States. Furthermore, the horse racing industry generates billions of dollars each year. The Association of Racing Commissioners International, Inc., located in Lexington, Kentucky, brings together the thoroughbred and harness racing industries. This self-contained group attempts to unify racing rules from state to state. They also hold an annual convention and ask for and provide constant feedback to state racing authorities.

SUMMARY

Most breeds of light horses are descendants from only a handful of original breeds. All draft horses descended from the Flemish horse. Horses are often bred to perform a certain task regardless of any purebred pedigree. Mares are seasonal breeders, normally conceiving only in the summer. Nutrition for horses relies on pasture (in season), good quality grass, or legume mixed hay and grain. The amount of feed and proportions of hay and grain required depend on the size of the horse and its activity level. Young growing horses and lactating mares have high nutrient requirements. Methods of horse identification, age determination, exercising, and hoof care are all management practices to be mastered by equestrians. Parasites infesting horses are similar to those of other livestock species. Vaccines are available for most horse diseases. Horses are housed in open sheds or in box stalls.

Over 9 million horses can be found across the United States. These horses are used primarily for pleasure although about one in five horses "work" for a living. Moreover, the horse industry directly accounts for over $39 billion of goods and services each year. A wide variety of horse-related careers are available for employment including trainers, grooms, stable managers, and race course personnel. Industry concerns center on unwanted horses, shortage of veterinarians, and animal identification. Many organizations support equestrians. Except for the AHC, these groups tend to be interest and event oriented.

CHAPTER REVIEW

EXPERIENTIAL LEARNING OPPORTUNITIES

1. Visit the American Veterinary Medical Association's (AVMA) Web site to investigate veterinary careers related to the equine industry.

2. Job shadow a local equine veterinarian.

3. Write a speech for an FFA competition on the pros and cons of horse slaughter for consumption.

4. Enter an FFA agriculture issues forum competition using National Animal Identification as a topic.

DEFINE ANY TEN KEY TERMS

AHC	equestrian
anthelmintic	farrier
band	founder
colic	hand mating
distemper	longeing
draft	sulky
dressage	tack
encephalomyelitis	

QUESTIONS AND PROBLEMS FOR DISCUSSION

1. Explain the origin of the name "Quarter Horse."

2. Over 75 percent of all horses carry this breed's blood. Name the breed.

3. To meet registration requirements, Paint Horses must trace their ancestry to the following three breeds: _____, _____, and _____.

4. Which breed is credited with being the first to be developed in the United States?

5. Are mares naturally receptive to stallions during the winter?

6. During the breeding season, mares normally cycle every _____ to _____ days.

7. Explain post-gastric fermentation.

8. Creep feeding in foals should begin at _____ to _____ weeks of age.

9. How many pounds of grain should a 400-pound, weaned foal consume?

10. List two methods of identifying horses.

11. Teeth angle sharply forward in young/old horses. Select one.

12. Founder is caused by _____.

13. Tetanus invades the body via _____.

14. Write a common box stall dimension.

15. Define tack.

16. Horses can be typed into three categories. Name them.

17. Eighty percent of horses in the United States are used for _____.

18. The horse industry directly accounts for _____ billion dollars each year in goods and services.

19. Concentrations of brood mares are located in this state. Name the state.

20. What is the color of Palomino Horses?

21. What is the average body temperature of a horse?

22. What do the letters AHC stand for?

23. True or False. Every horse in the United State must be identified by a traceable microchip.

24. How many United States slaughter houses kill horses for human consumption?

25. Name one association related to the rodeo industry.

26. Describe dressage.

27. The Association of Racing Commissioners International, Inc., is located in this Kentucky city. Name the city.

28. List two youth organizations that organize horse show events.

29. How many inches make a hand?

30. When was the Pony Express initiated?

Evaluation

Objectives

After completing this chapter, students should be able to:

▶ Explain the format of a judging contest

▶ Correctly complete a placing card

▶ Take notes for use in oral reasons

▶ Deliver a set of oral reasons

▶ Score an incorrectly placed class given an official placing and cuts

Key Terms

cull	grants	reasons
cut	keep–cull	transition phrase

Career Focus

A reporter asked Ellen, an accomplished defense attorney, which events in her life prepared her for a career in law. After some thought, Ellen replied that her experience as a member of her state-winning dairy-judging team was the single most important preparation she could have received. Ellen went on to elaborate that she was a very shy child. With the help of her judging-team coach, she became confident in her ability to note differences within groups of dairy animals. Further practice taught her to organize those differences into a set of reasons, and even more practice taught her to deliver those reasons verbally in a confident, cohesive format. Ellen said the skills she uses daily as a defense attorney were cultivated and honed as a judging-team member.

INTRODUCTION

Judging contests serve many purposes. Along with teaching students to evaluate animals or cuts of meat, these contests teach many important life skills. The decisions made during the judging contest require a contestant to draw from all the pertinent information available and use logical problem-solving skills. This

Delmar/Cengage Learning

FIGURE 13-1

Many youth take up judging in hopes of someday analyzing animals at a high-profile livestock show.

Animal Science Facts

In the 1920s, vocational agriculture students were invited to participate in live-stock judging contests at the American Royal Rodeo held annually in Kansas City, Missouri. That event was the precursor to the current National FFA livestock judging contest.

decision-making process must be clarified, organized, and defended when giving oral or written **reasons**, or explanation of the logic used for one's placing order. Finally, the oral delivery of reasons improves public-speaking skills, while written reasons help one's writing and organizational skills. Few students may evaluate livestock or meat after their scholastic days, but the skills and self-confidence gained in judging contests will remain with them for life.

CONTEST FORMAT

Judging contests are normally based on classes (groups) of four. These four animals, carcasses, or cuts of meat (exhibits) are presented in a group. Each exhibit is randomly identified by a number. Unless performance data indicating age are given, contestants have to assume that all animals in the class were all born the same day. Depending on the size of the contest and the amount of space available, half the classes may be displayed at once. For example, if the contest contains six classes, three may be displayed at one time. After the first three classes have been evaluated, the contestants take a break while the other three classes are prepared for viewing. If enough space is available, all classes can be presented at once. See figure 13-2.

Photo courtesy of Penn State Department of Dairy and Animal Science

FIGURE 13-2

Ideally, multiple classes can be displayed at once in a judging contest.

Upon arrival at a contest, judging-team members are divided into groups, usually designated by color or number, and are given a placing card for each class to be judged. If possible, team members are assigned to different groups. Each group will have a contest assistant acting as a group leader. When the contest begins, the group leader moves the group to the first class. Contestants are instructed to turn their backs to the class upon arrival. See figure 13-3. The group leader will announce when judging can begin.

The amount of time allowed for placing each class varies, but the standard is between 12 and 15 minutes. Contestants evaluate the class presented and rank the exhibits from best to poorest within the allotted time. For instance, if a contestant thinks exhibit number 4 is the best, number 2 the second best, number 3 the third best, and number 1 the poorest, the contestant's placing would be 4–2–3–1. If the contestants give reasons on the class, notes are taken to help contestants remember what the animals, carcasses, or cuts of meat look like.

At the end of the allotted evaluation period, time is called and contestants are instructed to turn their backs to the class. Each contestant then fills out a placing card and gives it to the group leader. The contestant's ranking of the class will be compared to an official placing, and a score is calculated. A perfect

Delmar/Cengage Learning.

FIGURE 13-3

Contestants begin the contest with their backs to the class.

score (correct placing) is worth 50 points. After cards are collected, the group moves to the next class.

Contestants are not allowed to talk, use reference materials, or electronic devices such as cell phones during the contest. Infractions of these rules will result in expulsion from the contest or a zero score for the class. Contestants routinely forget to record a placing or fill in their contestant numbers. Contest scorers have no way of knowing the card's owner or an intended placing, so a score of zero is awarded to contestants who improperly complete the placing card.

Placing cards come in several varieties. Manually scored cards contain a space for contestant number, class name, and four blanks for the contestants' placing of the four exhibits. Other placing cards contain the same blanks for contestant number and name of the class; however, instead of writing the placing, all possible placings of the class are listed on the card. The contestant must find the desired placing and simply mark that placing with an X. See figure 13-4.

Courtesy of Penn State Department of Dairy and Animal Science

Contestant Information		
CONTESTANT ID		
CLASS NAME		
CLASS NUMBER		
REASON SCORE		

1234		A
1243		B
1324		C
1342		D
1423		E
1432		F
2134		G
2143		H
2314		I
2341		J
2413		K
2431		L
3124		M
3142		N
3214		O
3241		P
3412		Q
3421		R
4123		S
4132		T
4213		U
4231		V
4312		W
4321		X

FIGURE 13-4

Sample placing card.

Similar cards may require the contestant to circle the desired placing or fill in a scan form bubble to indicate the desired placing.

Another type of class found in many livestock judging contests is known as a **keep–cull** class. In a keep–cull class, eight animals are presented instead of four. The objective of this class is to simulate decisions livestock producers make in selecting replacement breeding stock. The contestant must select four animals to keep as replacements and four animals to **cull** or eliminate or sell, often with the aid of performance data for the eight animals. Contestants mark the four numbers of the animals to keep or cull (depending on the contest—some contests require the four cull animals to be selected) on their placing cards. Again, a correct placing would result in 50 points.

While the contest is in progress, experienced official judges rank each class, take notes on the important characteristics of the class, and assign an official placing. At the end of the contest, these official placing and reasons will be shared with the participants.

REASONS AND QUESTIONS

Many contests, especially for experienced participants, require contestants to defend their placings through oral (livestock and dairy contests) or written (meat contests) reasons. Oral reasons are given to an official judge in a one-on-one setting. Notes taken during the class may be used as an aid in preparation for a set of reasons. The presentation of reasons must be made from memory. Accomplished judges give reasons while visualizing the animals in their mind's eyes. While this level of skill takes considerable practice, in the long term one can simply describe what is being seen instead of reciting a memorized speech.

There are four components of a good set of reasons: accuracy, organization, correct terminology, and delivery.

ACCURACY

Accuracy is the first and the most important component of a good set of reasons. Contestants should tell the truth. They should not try to invent reasons just to impress the judge. If there was no difference between two animals concerning a certain characteristic, it should not be discussed in the reasons. Gender differences among the animals in mixed-market classes should be noted and included in a set of reasons.

ORGANIZATION

Organization is the second component in a good set of reasons. A set of reasons is organized into three pairs. The top pair compares the first and second placings, the middle pair compares the second and third placings, and the bottom pair compares the third and fourth placings. The order of reasons always follows the sequence: top pair, middle pair, bottom pair. A **transition phrase** lets the judge know when the contestant is finished discussing one pair and is moving on to the next pair.

There is a logical order of discussion within each pair. In this example pair, 1 placed above 2.

1. The important reasons why 1 placed over 2. Major reasons are mentioned first in an opening statement, followed by smaller details that support the opening statement.

2. Any characteristics about 2 that were better than 1 (called **grants**).

3. Faults of 2 (optional).

The basic organization of an entire set of reasons follows:

1. Introduction: "I placed this class of vermites (an imaginary animal) 4–3–2–1." Note that numbers are used instead of names to identify the animal.

2. Opening statement, detailed reasons, grants, and faults for the top pair.

a. Opening statement: "In my top pair, 4 placed over 3 because 4 was bigger and had more desirable fur."

b. Detailed reasons: "Four was taller and had longer legs. In addition, 4 had longer, more luxurious fur, particularly between its ears and on its front legs."

c. Grants: "I grant that 3 had a longer, bushier tail."

d. Faults: "However, she had bad breath." Note that gender can be included as an identification point in place of the animal number.

3. Transition to middle pair: "Moving to my middle pair, I placed 3 over 2."

4. Opening statement, detailed reasons, grants, and faults for middle pair.

5. Transition to bottom pair.

6. Opening statement, detailed reasons, and grants for bottom pair.

7. Reasons why 1 was last. "Nonetheless, 1 was the skinniest, smelliest, most uncoordinated vermite in the class."

8. Conclusion: "Thank you."

Obviously, the terminology used for an actual set of reasons would be appropriate for the species in question.

CORRECT TERMINOLOGY

Correct terminology is the third component of a good set of reasons. Use of incorrect terminology quickly identifies the contestant as a novice. For example, muscling in the rear leg of a pig is called the ham. The same anatomical location of a lamb is called the leg, and in a steer is known as the quarter.

DELIVERY

The final component in a set of reasons is the delivery. The contestant should enter the reasons room confidently, remain 8 to 12 feet from the judge, and stand

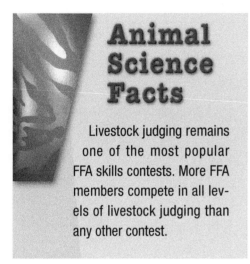

Animal Science Facts

Livestock judging remains one of the most popular FFA skills contests. More FFA members compete in all levels of livestock judging than any other contest.

with feet squarely placed and hands held behind the back. See figure 13-5. Hats and notes should be left outside the reasons room. Reasons should be delivered in a slow, smooth, confident, and authoritative manner. The tone should be conversational, be loud enough to be understood, and show good voice inflection. Good eye contact with the judge demonstrates confidence. Practicing reasons in front of a mirror can help in gaining confidence and improving eye contact.

A well-delivered and organized set of reasons should not last more than two minutes. In some contests contestants may be cut off if reasons exceed the two-minute time limit.

Taking notes serves to remind the participant of important physical characteristics that influenced class placing. Logical arrangement of notes eases the organization process in preparing a set of reasons. A sample note-taking format is shown in figure 13-6, and the sample notes for the class of vermites is shown in figure 13-7. Judges rely less and less on written notes as they gain more experience. After much practice, classes of livestock seem to imprint

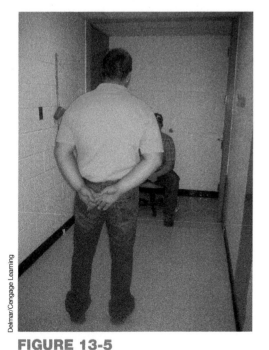

FIGURE 13-5

Ideal reasons stance.

FIGURE 13-6

Organized note-taking format for a judging class of four exhibits.

Placing 4 3 2 1

Class Name: Vermites	Class Number: 1	Official _____
$^4/_3$ Bigger — long legs — taller Better fur — longer — quality	$^3/_4$ Long Bush Tail	$^3/$ Bad Breath
$^3/_2$ Movement — quicker — takes longer steps Sharper Teeth	$^2/_3$ Pointier Ears	Limps
$^2/_1$ Longer eyelashes — more refined Toes — Longer — more — Better pedicure	$^1/_2$ Longer tongue	$^1/$ Skinny Smelly Uncoordinated

Delmar/Cengage Learning

FIGURE 13-7

Sample notes for a class of fictional animals.

themselves in the mind's eye, where they can be called upon and accurately described during a set of reasons.

Younger participants in a judging contest may be asked questions about a class of exhibits in a question session. An example question follows: "Between 1 and 4, which vermite had longer whiskers?" Questions are answered on a score card, without the use of notes, so the ability to visualize the animals aids in scoring well in question sessions. See figure 13-8.

		QUESTIONS CLASS						

CONTESTANT ID _____

CLASS NAME _____

Circle your answer ⟶ ○

	1	2	3	4	5	T	F	
1.	1	2	3	4	5	T	F	____
2.	1	2	3	4	5	T	F	____
3.	1	2	3	4	5	T	F	____
4.	1	2	3	4	5	T	F	____
5.	1	2	3	4	5	T	F	____
6.	1	2	3	4	5	T	F	____
7.	1	2	3	4	5	T	F	____
8.	1	2	3	4	5	T	F	____
9.	1	2	3	4	5	T	F	____
10.	1	2	3	4	5	T	F	____

____ Score

Photo courtesy of Penn State Department of Dairy and Animal Science

FIGURE 13-8

Sample questions card.

CUTS AND SCORING

Between each pair in the placing, officials assign a **cut**. Each cut is a number of points that indicate the difficulty of that placing. For instance, if the top pair was very close (a difficult decision), the cut would be small (one or two points). A large cut would be assigned when one animal of the pair easily places higher than the other animal. Six or seven points would be a large cut. Placings of intermediate ease receive intermediate cuts. Total cuts cannot exceed 15.

Cuts are used to score incorrectly placed classes. To illustrate, the official placing of a class was 4–3–2–1, with cuts of 2–5–4. The contestant placed the class 3–4–2–1. The contestant switched the top pair and lost the top pair cut (two points). Thus, the contestant's score would be 50 − 2 = 48. A middle pair switch (4–2–3–1 placing) would score 45 (50 − 5 = 45), while a bottom pair switch (4–3–1–2 placing) would score 46 (50 − 4 = 46). A correct placing would receive 50 points.

QUESTION	IF YES	IF NO
1. Was 1 placed over 2?	Subtract 0	Subtract a
2. Was 1 placed over 2?	Subtract 0	Subtract a + b
3. Was 1 placed over 4?	Subtract 0	Subtract a + b + c
4. Was 2 placed over 3?	Subtract 0	Subtract b
5. Was 2 placed over 4?	Subtract 0	Subtract b + c
6. Was 3 placed over 4?	Subtract 0	Subtract c

Use this formula to compute scores after a contest. Fortunately, contest officials have computer programs and precalculated scoring grids to help score contests.

FIGURE 13-9

How to score a judging class. Official placing 1-2-3-4 with cuts of a, b, c.

Other incorrect placings are more difficult to score. Every possible placing must be analyzed and the proper cuts deducted. In all situations, the process seen in figure 13-9 must be followed.

OTHER TYPES OF CONTESTS

Although judging contests are the most common animal-based events, other contest types are offered at various venues. Livestock evaluation contests test contestants' abilities to estimate slaughter animal carcass composition and ultimate value in the marketplace. The National Western Livestock Show in Denver, Colorado, offers a carload competition where teams compete in their abilities to select groups of beef cattle rather than individual animals. Various quiz bowl contests pit teams against one another to determine industry and production knowledge. Skillathon or Stockmen's competitions are typically individual events that test competitors' knowledge on a host of topics from feeds to equipment to anatomy. See figure 13-10.

Delmar/Cengage Learning

Photo courtesy of Melanie Barkley, Bedford County, Penn State Cooperative Extension Service

FIGURE 13-10

Skillathon contests test student knowledge of many facets of the livestock industry.

SUMMARY

Judging contests assist students in developing critical thinking, organizational, and speaking skills. Along with placing classes, students are asked to justify the placing with reasons. The four key components of a good set of reasons include: accuracy, organization, correct terminology, and delivery. Accuracy is the most important component. Keep–cull classes require students to select animals to be kept as herd replacements. Classes are scored with a system of cuts that more harshly penalize obvious pair switches. Other types of contests test knowledge of the breadth of animal industries.

CHAPTER REVIEW

EXPERIENTIAL LEARNING OPPORTUNITIES

1. Sponsor a skillathon contest for 4-H and FFA livestock or dairy exhibitors in your area. Have students plan and execute the event.

2. Help score contestants' placings at a livestock or dairy judging contest.

3. Practice saying a short memorized speech (the FFA creed, e.g.) using the correct reasons stance, volume, inflection, and eye contact.

DEFINE ALL KEY TERMS

cull

cut

grants

keep–cull

reasons

transition phrase

QUESTIONS AND PROBLEMS FOR DISCUSSION

1. Judging classes usually contain _____ (number) animals, carcasses, or cuts of meat (exhibits) to place.

2. Animals, carcasses, or cuts are identified by a _____ within each class.

3. A perfect score for a correct placing earns the contestant _____ points in a judging contest.

4. True or False. Judging contests normally begin with participants facing the classes.

5. Contestants in judging contests can earn a zero score for several infractions. Name three of them.

6. Most judging classes last for _____ to _____ minutes.

7. A _____ class simulates producer selection of herd replacements.

8. Are judging contest participants allowed to take notes during the classes?

9. Explain the four components of a good set of reasons.

10. Define cull.

11. What is the purpose of a keep–cull class in a livestock judging contest?

12. What is a grant?

13. Describe the stance competitors should take when delivering oral reasons.

14. How far should a contestant stand from the judge when giving reasons?

15. How much time does a contestant have to give a set of oral reasons?

16. If a contestant switched the top pair in class with cuts of 6–2–1, what score would the contestant earn?

17. What FFA skills contest remains the most popular?

18. Describe the origin of the National FFA Livestock Judging Contest.

19. Which western city mentioned in the text hosts a carload beef cattle judging contest?

20. True or False. Written reasons may be required in some meats judging contests.

Objectives

After completing this chapter, students should be able to:

▶ Explain the significance of performance data to livestock breeders

▶ Compare EPDs, EBVs, multiple trait indexes, actual data, and ratios

▶ Discuss why selection may vary depending on environmental conditions

▶ Use performance data in judging classes and on-farm selection

Key Terms

average daily gain (ADG)

backfat at 250 pounds (BF)

birth weight EPD (BW EPD)

days to 250 pounds (DAYS)

estimated breeding value (EBV)

expected progeny difference (EPD)

flock EPDs or FEPDs

general purpose index (GPI)

individual data

lifetime net merit (LNM)

litter weight at 21 days (LW21)

maternal line index (MLI)

maternal milk (MM EPD)

multiple trait indexes

number born alive (NBA)

percent difficult births in heifers (%DBH)

predicted transmitting ability (PTA)

predicted transmitting ability for type (PTAT)

production type index (PTI)

PTA$ for cheese yield (CY$)

PTA$ for milk and fat (MF$)

PTA$ for milk, fat, and protein (MFP$)

ratio

scenario

sow productivity index (SPI)

terminal sire indexes (TSI)

type-production index (TPI)

weaning weight EPD (WW EPD)

yearling weight EPD (YW EPD)

Career Focus

Korry raises Boer goats in his small Oregon backyard. His goal is to produce meat goats that grow quickly and efficiently. Korry is envious of the myriad of performance data available to his classmates who raise pigs, beef, and dairy cattle because they all have a variety of readily available information on which to base their breeding stock selections. Unfortunately, that type of information is not available to goat breeders. Someday, Korry thinks, he will study to become an animal geneticist and work to decipher the goat genome to discover the genes responsible for goat growth and efficiency.

INTRODUCTION

Livestock and most livestock products are sold by the pound. However, individual animals within a species differ in their genetic ability to produce product. Livestock producers use performance data to identify animals that carry desired genes for growth, leanness, milk production, or other economically important traits. Animals with the best performance data can then be retained as sires and dams for the next generation. Rapid genetic improvement on any trait relies heavily on the use of performance data.

TYPES OF PERFORMANCE DATA

Traditional performance data can be divided into four general categories. Individual data, ratios, and indexes are relatively simple. The fourth category includes more complicated genetic predictors such as estimated breeding values, expected progeny differences, and predicted transmitting abilities.

INDIVIDUAL DATA

Individual data reflect the performance of an individual animal for a specific trait. For example, a bull's individual performance for **average daily gain (ADG)**

Delmar/Cengage Learning

FIGURE 14-1

For producers interested in performance data, scales are a valuable tool.

may be 4.2 pounds per day. Individual data form the bedrock upon which all other performance data categories are based.

Calculation of individual data is based on accurate records. Three pieces of information needed to calculate the above bull's average daily gain are: First, a beginning, or on-test weight at the start of the feeding period; second, an ending, or off-test weight at the conclusion of the feeding period; and third, the length of the feeding period in days. Average daily gain is calculated using the formula:

(Off-test weight – on-test weight)/Days on feed

Other individual data, such as birth weight, number born, number weaned, backfat depth, loineye size, weaning weight, and milk production, require only one measurement. See figure 14-2.

RATIOS

Other types of performance data require individual measurements from more than one animal. Individual data from a contemporary group of animals allow comparison within the group by using a **ratio**. Ratios compare animals using a percentage approach for a single trait. Using the bull example mentioned earlier, if the average of the bull's contemporary group was 4.2 pounds of gain per day, the previously mentioned

Photo by Frank Flanders

FIGURE 14-2

This bulk tank stores and cools milk that is sold by the pound.

BULL NUMBER	AVERAGE DAILY GAIN (LBS/DAY)	RATIO
1	4.2	100
2	4.0	95
3	4.5	107
4	4.1	98
Average	4.2	100

FIGURE 14-3

Bull ratios for average daily gain.

bull's ratio would be 100 (average of the contemporary group). Most ratios (where higher numbers are more desirable) are calculated using the formula:

Individual animal's measurement/contemporary group average measurement × 100

Another bull from the same contemporary group with an average daily gain of 4.0 pounds per day would have a ratio of 95 (5% below average). A bull that gained 4.5 pounds per day would have a ratio of 107 (7% above average). See figure 14-3.

INDEXES

Geneticists can formulate indexes to select for several traits simultaneously. These indexes are called **multiple trait indexes**. For example, terminal sire boars may be selected for average daily gain, backfat, and loineye size at the same time using a multiple trait, terminal sire index. A terminal index would compare all three measurements to the contemporary average and assign a weight to each trait according to its economic importance. For instance, if a 1 percent change in backfat is worth $0.10, and a 1 percent change in loineye size is worth $0.05, backfat would be weighted twice as heavily as loineye size. Like single trait ratios, multiple trait indexes are based on 100. Maternal indexes can be calculated emphasizing traits such as weaning weight and milk production with a lesser emphasis on carcass traits. Ratios and indexes more accurately show genetic value than individual data because of the comparison with contemporaries.

EBVs, EPDs, AND PTAs

An **estimated breeding value (EBV)** takes the index concept one step further. EBVs include the heritability of a trait. Also, EBVs present an estimate of the animal's genetic worth as a parent in comparison to other animals in a group. EBVs use all available records of relatives within the same herd or flock. EBVs are presented in index form with 100 as an average. An EBV of 102 means that the animal's progeny should perform 2 percent above the average for that flock or herd for a specific trait. The accuracy value of an EBV increases as more records are added. Estimated breeding values are more useful than either indexes or individual data because records of related animals are incorporated into the calculation. See figure 14-4.

The most useful performance data for meat animal species are **expected progeny differences (EPD)**. Like EBVs, EPDs also include all available records of relatives and incorporate heritability. However, through the use of the best linear unbiased predictor (BLUP) computer program, EPDs can also include records from relatives outside the herd or of dead ancestors.

EPDs are presented in practical units such as pounds, days, or inches. For example, an Angus bull with an EPD for a weaning weight (WW EPD) of +10 pounds should, on average, sire calves with weaning weights 10 pounds heavier than those sired by an Angus bull with a WW EPD of 0. In contrast, an Angus bull with a –10 EPD for weaning weight should, on average, sire calves with weaning weights 10 pounds lighter than those sired by an Angus bull with a WW EPD of

COW ID A0021

PERFORMANCE TRAIT	ESTIMATED BREEDING VALUE	ACCURACY
Birth Weight	95	47%
Weaning Weight	102	35%
Maternal Milk	96	22%
Yearling Weight	103	25%

FIGURE 14-4

Estimated breeding values for beef cow A0021.

Delmar/Cengage Learning

0 pounds. If presented with individual data, indexes or EBVs, and EPDs, EPDs should be weighted more heavily during the selection process. EPDs can be used to compare animals of the same breed in different herds. See figure 14-5. The breed average EPD for a given trait is not necessarily zero. For example, the average **weaning weight EPD** (genetic difference in weaning weights of the bull's calves) for Angus bulls is about 45 pounds.

Dairy geneticists have improved on the EPD concept. In the dairy industry, EPDs are known as **predicted transmitting ability (PTA)**. Using current market prices for milk, percent butterfat, and percent protein, PTAs are translated into a dollar value that predicts the added income to be gleaned from an individual sire's daughters. For example, a bull with a **PTA for milk, fat, and protein (MFP$)** of $100 would sire daughters whose milk (because of increased milk production, fat, and protein content) would be worth $100 more per year than daughters of a bull with an MFP$ of $0. Again, these dollar values (or the raw PTAs) can be used only to compare sires within the same breed. PTAs can also be negative. Some beef breeds offer similar economic index tools for breeding stock selection.

The dairy industry combines many economically important traits including health traits, production

HEREFORD BULL CALVES

BULL NUMBER	DOB	BW EPD	WW EPD	YW EPD	MM EPD
1	2/1/07	5.3 (.25)	54 (.20)	89 (.20)	16 (.15)
2	1/16/07	5.0 (.21)	58 (.19)	85 (.19)	24 (.19)
3	1/24/07	5.6 (.35)	60 (.22)	95 (.22)	18 (.09)
4	1/9/07	3.7 (.30)	57 (.20)	86 (.20)	25 (.10)
Breed Avg.		3.7	46	56	15

DOB = Date of Birth   YW EPD = Yearling Weight EPD

BW EPD = Birth Weight EPD   MM EPD = Maternal Milk EPD

WW EPD = Weaning Weight EPD   ( ) = EPD Accuracy

FIGURE 14-5

Birth weight, weaning weight, yearling weight, and milk EPDs for a group of Hereford bull calves.

Delmar/Cengage Learning

traits, and conformation into a single selection index called **lifetime net merit (LNM)**. LNM estimates a diary cow's ability to return a profit over its lifetime.

The numbers generated by a BLUP are only as accurate as the number of records incorporated into the program. If few records are available, the accuracy (abbreviated acc., reported as reliability in dairy) is low. As more records are added to the animal's performance profile, accuracy of the BLUP increases. Increasing accuracy means the estimate of a trait by performance data is a true genetic estimate, and not a result of environmental conditions. EPDs are usually presented with a measure of the accuracy between 0 and 0.99. Dairy PTAs and EBVs remove the decimal points and provide a reliability score between 0 and 99. The closer the accuracy (reliability) number is to 0.99 (99), the more accurate the genetic estimate. Compare the following two beef bulls.

	WEANING WEIGHT EPD	(ACC.)
Bull A	+10	(0.15)
Bull B	+10	(0.89)

Although the weaning weight EPDs are the same, the accuracy of Bull B is much higher. More emphasis should be placed on the weaning weight EPD of Bull B. Chances are that Bull B is an older sire with many progeny records upon which to base the EPD. Bull A is most likely a younger sire whose EPD is primarily based on records of its relatives. As more records are added to Bull A's performance profile, his weaning weight EPD accuracy will increase. A good chance exists that Bull A's EPD will move up or down dramatically as the accuracy value improves. Bull B's EPD would be much less likely to change as more records are added.

GENOMICS

As mentioned in Chapter 6, genomics is the science of looking at an animal's genetic code to select animals with desired genes for certain traits. In the beef industry, genes that influence marbling, tenderness, quality

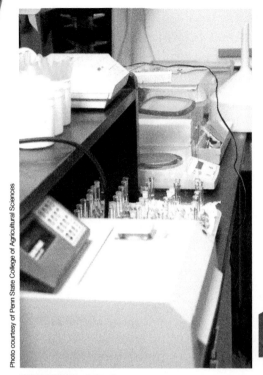

FIGURE 14-6

Genomics is a high-tech approach to genetic selection.

grade, and feed efficiency have been identified. In the dairy industry, scientists can analyze thousands of genes to look for patterns of genetic markers found in high-producing animals, then make selections for LNM based on the genetic screening. Reliability of these tests approaches 60 to 70 percent for dairy bulls, or about the same as if the prospective sire had 50 daughters in production. See figure 14-6. Thus DNA from newborn dairy bull calves can be quickly screened to determine if the genetic markers make them potential sire candidates. Those with poor genetic profiles can be castrated and those that are retained can be used with confidence as yearling sires.

DATA FOR INDIVIDUAL ANIMAL INDUSTRIES

The performance data presented to contestants in judging contests vary by species. Following is a discussion of the types of data likely to be encountered for each species.

SWINE

Individual data encountered may include birth date, **number born alive (NBA)** in the litter, number weaned in the litter, **litter weight at 21 days (LW21)**, average daily gain (ADG), **days to 250 pounds (DAYS)**, and **backfat at 250 pounds (BF)**. For the first four types of individual data, higher numbers are desirable. Days to 250 and backfat at 250 are two traits for which lower numbers are more desirable. Contemporary ratios may be calculated for any of the above traits.

Multiple trait indexes are also used in swine selection. **Sow productivity index (SPI)** combines NBA and LW21 to gauge maternal performance. **Maternal line index (MLI)** emphasizes maternal records with minor emphasis on growth and carcass records. **General purpose index (GPI)** gives equal weight to maternal, growth, and carcass traits. **Terminal sire indexes (TSI)** use days to 250 and backfat at 250 pounds. See figure 14-7.

EPDs for swine include NBA and LW21 for which higher EPDs are more desirable. Lower

BOAR ID 47-4 PERFORMANCE RECORD

INDIVIDUAL TRAITS	EPD	INDEX TRAITS	INDEX
Number Born Alive	+0.08	Maternal Line	103.2
21-day litter weight	+1.25	Terminal Sire	97.3
Days to 250	−0.90	Sow Productivity	102.3
Backfat thickness	−0.03		

FIGURE 14-7

Boar 47-4's performance record. He exhibits strong maternal traits.

numbers are preferable for days to 250 and backfat at 250 pounds.

BEEF

Actual data encountered in beef performance classes include birth date, birth weight, weaning weight, yearling weight, frame score or hip height, and the scrotal circumference for bulls. Contemporary ratios of all these traits may also be encountered.

EPDs for beef cattle include **birth weight EPD (BW EPD)** (lower numbers mean lower birth weights), weaning weight EPD (WW EPD), **yearling weight EPD (YW EPD)**, and **maternal milk (MM EPD)**, the difference in pounds of calf weaned due to milk production. EPDs for carcass weight, external fat thickness, ribeye area, marbling score (intramuscular fat), and percent retail product are also available for certain beef breeds. See figure 14-8.

CARCASS EPD DATA – MAINE ANJOU BULLS

BULL NUMBER	CARCASS WEIGHT	% RETAIL PRODUCT	MARBLING	FAT THICKNESS	RIBEYE AREA
1	+22.2 (.17)	+0.43 (.17)	+0.04 (.20)	−0.03 (.29)	+0.52 (.19)
2	+4.9 (.36)	+0.42 (.36)	+0.03 (.34)	−0.03 (.34)	+0.30 (.29)
3	+10.6 (.11)	+0.55 (.11)	+0.25 (.14)	+0.01 (.19)	+0.77 (.13)
4	+14.8 (.30)	+0.00 (.30)	+0.55 (.30)	−0.02 (.34)	+0.40 (.26)
Breed Avg.	+2.5	+0.30	+0.21	+0.00	+0.16

( ) = EPD Accuracy

FIGURE 14-8

Sample carcass EPD data for a set of Maine-Anjou bulls.

Animal Science Facts

In the past, it was necessary to slaughter an animal to collect data related to carcass traits such as backfat and ribeye area. Obviously it is impossible to use a carcass as breeding stock if it did turn out to be exceptionally lean or heavily muscled. Until the advent of real-time ultrasound these data had to be collected from close relatives and progeny and results used to discern the genetic potential of the animal in question. Real-time ultrasound allows seed stock producers to see and measure the actual amount of external fat, intramuscular fat, and muscle in a live animal. Many potential sires and dams can be ultrasonically scanned and only the best retained as parents. A real-time ultrasound probe emits sound waves through the skin, fat, muscle, and bone. Sound waves reflect from any change in tissue type. The reflected sound waves are collected and processed by a sophisticated machine and the resulting cross section is displayed on a screen. Actual measurements can then be made from this realistic picture. Beef cattle EPDs for external fat, intramuscular fat (marbling), and muscling have evolved because of the use of ultrasound.

Economic indexes such as weaned calf value ($W), cow energy requirements ($EN), feedlot performance ($F), carcass grid value ($G), and feedlot and carcass grid value combined ($B) allow selection for multiple traits that influence profitability.

SHEEP

Individual data for sheep are much the same as other species. Birth date, birth weight, type of birth and/or rearing (Single = S, Twin = TW, Triplet = TR, Quad = Q), weaning weight (adjusted to 30, 60, 90, or 120 days), and yearling weight are all data that could be included in a judging class. Data for classes of wool breeds may include grease fleece weight (weight of unclean fleece), clean fleece weight, staple length, or grade.

Indexes can be calculated for the earlier mentioned data except the birth date.

EPDs are available for sheep, but too little data are available for most breeds to compare EPDs with animals outside the flock. Therefore, EPDs for most breeds of sheep must be calculated within a flock and are called **flock EPDs** or **FEPDs**. Sheep FEPDs are available for number of lambs born, percent lamb crop (number of lambs born per 100 ewes), weaning weight (adjusted to a constant weight), fleece weight, staple length, and fiber diameter. See figure 14-9.

DAIRY

Individual dairy production data are available for milking records and type. Pounds of milk, protein, and fat per lactation or lifetime are common individual production records. Average percent protein and fat in the milk are also frequently used. Type scores include a variety of visual evaluations including udder quality, feet and legs, dairy character, and body depth. PTAs are also available for fertility and longevity. Calving ease is reported as **percent difficult births in heifers (%DBH)**. This number is an estimate of the number of calving problems expected per 100 heifers bred to a specific bull.

Type-production index (TPI) combines production with type into a general purpose index for Holsteins. Other breeds use a similar index with a similar name: **production type index (PTI)**. Both indexes use 100 as average.

In addition to LNM, common individual PTA values available for dairy cattle include pounds of fat, percent fat, pounds of protein, percent protein, and pounds of milk. PTA$ data are available for **milk, fat, and protein combined (MFP$)**. Portions of MFP$ can be isolated depending on production emphasis. **PTA$ for cheese yield (CY$)** predicts the value of milk protein content. **PTA$ for milk and fat (MF$)** predict the economic difference due to milk production and butterfat content. **Predicted transmitting ability for type (PTAT)** indicates the number of type points that will be gained or lost compared with the breed average for a specific sire's daughters. PTAT is the only performance data that are subjective in nature. See figure 14-10 for dairy performance data.

DORSET RAM FLOCK EPDs

RAM NUMBER	BIRTH TYPE	PERCENT LAMB CROP FEPD	60-DAY ADJ. WEIGHT FEPD	FLEECE WEIGHT FEPD
1	Twin	–2.2 (.25)	–1.1 (.22)	+0.5 (.15)
2	Twin	+3.6 (.19)	+1.7 (.19)	–0.3 (.15)
3	Single	+8.3 (.21)	+0.3 (.21)	–0.4 (.15)
4	Twin	+1.1 (.19)	+0.1 (.15)	–0.1 (.15)

FIGURE 14-9

Example Flock EPDs for four Dorset rams.

DAIRY BULL PERFORMANCE DATA

BULL ID	LNM$	PROTEIN	FAT	PTA TYPE	%DBH
Homer	557 (79)	40 (81)	64 (80)	+0.9 (78)	7 (80)
Gomer	225 (87)	10 (89)	24 (88)	+2.4 (87)	5 (89)
Boomer	865 (57)	57 (60)	101 (61)	+2.2 (55)	10 (58)
Roamer	458 (99)	36 (99)	60 (99)	+1.9 (99)	8 (99)

LNM$ = Lifetime Net Merit

Protein = Pounds of protein

Fat = Pounds of Fat

PTA Type = Predicted transmitting ability for type

% DBH = Percent difficult births if bred to heifers

( ) = reliability or accuracy

FIGURE 14-10

This is a small smattering of the performance data types available for dairy bulls.

Animal Science Facts

The American Angus Association recently introduced an EPD for docility. Not only are docile beef cattle easier to work with and less dangerous to handlers, but there is a high positive correlation between docility and feedlot feed efficiency.

USES OF DATA IN PRODUCTION SITUATIONS

Performance data are used to predict the genetic differences among animals. Producers and judges identify the genetic traits important in a given production situation. Students must learn to analyze production situations, identify the genetic traits that should be emphasized in that situation, and then select livestock based on the necessary emphasis.

Judging contests that include performance data for a class normally include a **scenario**. Scenarios inform the judges about the production situation in which the animals will be placed. The scenario influences how the judge analyzes the performance data and phenotypic traits. A properly constructed scenario should contain the following information:

1. Use of animals in a pure or crossbreeding program

2. Availability of resources and environment for the herd or flock

3. Marketing strategy for the progeny

The scenario gives clues as to which data should be emphasized in selection. Beef cattle perhaps offer the best illustration of interpreting scenarios. Other livestock species require similar thought processes. Following are two beef cattle scenarios accompanied by an analysis. See figures 14-11 and 14-12.

© Justin Mair. Used under license from Shutterstock.com

FIGURE 14-11

Beef cattle raised in dry range conditions require performance data to fit their environment.

© Christopher Elwell. Used under license from Shutterstock.com

FIGURE 14-12

Producers raising beef cattle in Eastern or Midwestern environments may have different performance data criteria than those in areas with fewer feed resources.

Scenerio

SCENARIO 1: LIMOUSIN BULLS

(1) Bulls will be bred to registered mature cows. (2) Feed and labor resources are adequate and typical of those on a small, Northeastern farm. (3) Some heifer calves will be retained as replacements. (4) All bull calves will be retained, performance tested, and sold as commercial bulls to producers selling slaughter cattle on a carcass grid that rewards high-quality, high-cutability cattle.

Analysis

1. ***Bulls will be bred to registered mature cows.***

   When bulls will be bred to heifers, birth weight data will be a concern. Heifers are prone to dystocia if calves are too big. Birth weight EPD predicts the size of the calves sired by a bull. Since bulls are bred to mature cows, higher birth weight EPDs could possibly be tolerated.

2. ***Feed and labor resources are adequate and typical of those on a small, Northeastern farm.***

   Eastern and Midwest producers often have access to higher-quality forages and more available labor at calving time than producers in range states. Therefore, larger-framed, later-maturing, higher birth weight EPD bulls can be tolerated than if feed and labor resources were typical of those on the Western range.

   (continues)

3. *Some heifer calves will be retained as replacements.*

Since some daughters of these bulls will be retained as replacements, milk production and weaning weight EPDs should be considered. Those retained heifers will be expected to have their own calves someday, so birth weight EPDs should not be excessively high in comparison with other bulls.

4. *All bull calves will be retained, performance tested, and sold as commercial bulls to producers selling slaughter cattle on a carcass grid that rewards high-quality, high-cutability cattle.*

The final sentence is perhaps the most important one in this scenario. Bulls used to sire calves sold under a grade and yield program must be fast-growing (high weaning and yearling weight EPDs), and possess strong EPDs for carcass traits.

The overall evaluation of this scenario should lead the judge to this conclusion: the ideal bull for this scenario should be fast-growing, should be heavily muscled, and have adequate milk production EPDs. Birth weight EPDs should not be excessively high in comparison with other bulls in the class.

SCENARIO 2: ANGUS HEIFERS

(1) These heifers will be naturally mated to Simmental bulls. (2) Female offspring will be retained as commercial females. (3) All male offspring will be castrated and sold as feeder steers. (4) These heifers and their progeny will be raised on range conditions where feed and labor resources are limited.

Analysis

1. *These heifers will be naturally mated to Simmental bulls.*

As a breed, Simmentals are not known for their calving ease but are heavily muscled and fast growing. Therefore, birth weight EPDs should be emphasized over growth EPDs for the Angus heifers.

2. *Female offspring will be retained as commercial females.*

The Angus $\times$ Simmental female progeny will be used as commercial cows, so emphasis should be on maternal EPDs.

3. *All male offspring will be castrated and sold as feeder steers.*

A significant proportion of income will come from the sale of feeder steers. The heavier the steer calves at weaning, the bigger the producer's paycheck; so weaning weight EPDs are important.

4. *These heifers and their progeny will be raised on range conditions where feed and labor resources may be limited.*

Since feed and labor may be limited, low-maintenance, easy-keeping heifers should be emphasized. Limited feed resources suggest that heifers with excessively high milk production EPDs should be avoided. The nutrient demands of high milk production combined with limited feed can lead to thin heifers that will not rebreed. Again, low birth weight EPDs are emphasized because of the lack of labor at calving time. Heifers must be able to calve unassisted. If $EN values are available, these can be used to select energy efficient heifers to fit the scenario.

Evaluation of this scenario points toward the selection of Angus heifers that have low birth weight EPDs, moderate milk production EPDs, and high weaning weight EPDs. Yearling weight EPDs should receive the lowest consideration.

SUMMARY

Several types of performance data are available to aid in the selection of breeding livestock. Actual data are the easiest to gather, but the least accurate. Ratios and multiple trait indexes compare animals within a contemporary group. EBVs, EPDs, and PTAs offer a more accurate assessment of an animal's genetic potential because they include individual data from the animal and related family members. The accuracy or reliability of EPDs and PTAs increases with the number of records available. Genomics allows producers to short-circuit the selection process by analyzing an animal's genome for desired traits early in life. When using performance data for the selection of livestock, the judge must consider the environment and intended use of the animals in a crossbreeding scheme. This information is presented to judging contest contestants in the form of a scenario for each performance class. Critical thinking skills must be used to analyze scenarios and select the animals with the correct balance and fit of performance data. Students should recognize that specific performance data traits and indexes evolve continuously as geneticists and economists uncover new information.

CHAPTER REVIEW

EXPERIENTIAL LEARNING OPPORTUNITIES

1. Contact a local agricultural educator or extension educator for information on judging contests with performance data. Participate in one.

2. Job shadow an employee of the Dairy Herd Improvement Association (DHIA)

DEFINE ANY TEN KEY TERMS

average daily gain (ADG)

backfat at 250 pounds (BF)

birth weight EPD (BW EPD)

days to 250 pounds (DAYS)

estimated breeding value (EBV)

expected progeny difference (EPD)

flock EPDs or FEPDs

general purpose index (GPI)

individual data

lifetime net merit (LNM)

litter weight at 21 days (LW21)

maternal line index (MLI)

maternal milk (MM EPD)

multiple trait indexes

number born alive (NBA)

percent difficult births in heifers (%DBH)

predicted transmitting ability (PTA)

predicted transmitting ability for type
(PTAT)

production type index (PTI)

PTA$ for cheese yield (CY$)

PTA$ for milk and fat (MF$)

PTA$ for milk, fat, and protein (MFP$)

ratio

scenario

sow productivity index (SPI)

terminal sire indexes (TSI)

type-production index (TPI)

weaning weight EPD (WW EPD)

yearling weight EPD (YW EPD)

QUESTIONS AND PROBLEMS FOR DISCUSSION

1. Give an example of individual data that require three pieces of information.

2. List five types of individual data that use only one measurement.

3. A ratio compares animals within the same _____ group.

4. True or False. Individual data form the basis for other types of performance data.

5. The average ratio score is _____.

6. _____ presents an estimate of the animal's genetic worth as a parent in comparison to others in the same herd.

7. _____ presents an estimate of the animal's genetic worth as a parent when compared with animals inside and outside the herd.

8. True or False. EPDs are presented as a number with 100 being the average score.

9. True or False. The BLUP allows for use of data from dead ancestors in EPD calculations.

10. The _____ industry uses predicted transmitting ability to forecast the added income to be gained from an individual's daughter.

11. Explain why EPDs and PTAs become more accurate as more individual records are used in computation.

12. Given individual data, ratios, EBVs, and EPDs, which should be given the highest consideration in selection? Why?

13. How can a scenario help a judging contestant consider performance data?

14. List four types of performance data commonly used in the swine industry.

15. List four types of performance data commonly used in the beef industry.

16. List four types of individual data commonly used in the sheep industry.

17. List three types of performance data commonly used in the dairy industry.

18. True or False. It is impossible to measure ribeye area on a live animal.

19. True of False. An EPD exists for docility in Angus cattle.

20. True of False. Simmental cattle are known for their growth rate and muscle volume.

Objectives

After completing this chapter, students should be able to:

▶ Define meat quality in two different ways

▶ Sketch beef, pork, and lamb carcasses and label the primal cuts

▶ Contrast ideal cooking requirements and conditions for roasts and steaks

▶ Differentiate between yield and quality grades

▶ Place classes of beef, pork, and lamb carcasses

Key Terms

break joint

buttons

choice/select spread

cutability

Institutional Meat Purchasing Specifications (IMPS)

kidney, pelvic, and heart fat (KPH)

lean-to-fat ratio

preliminary yield grade (PYG)

primal cuts

quality grades

spool joint

subprimal cuts

USDA meat graders

USDA meat inspector

yield grades

Career Focus

Callie knew she was fortunate to have been on a meats judging team while in FFA. Not many students had that opportunity ever, let alone prior to college. During her training, she learned the wholesale and retail cuts for beef, pork, and lamb. Callie did not go on to judge meats during college but pursued a degree in food science. Her first job after college was in the quality control department for a food distribution company. However, Callie found herself gravitating to the meats side of the food distribution business. Within two years, Callie was putting her high school meats training to good use by purchasing trailer loads of meat for the company's customers each day.

Delmar/Cengage Learning

FIGURE 15-1

Evaluating carcasses based on economic value takes practice.

INTRODUCTION

Live market livestock are visually evaluated to estimate the contents of the final carcass, namely meat. A basic knowledge of meat evaluation helps livestock judges better select animals that, when slaughtered, will produce carcasses containing large amounts of high-quality, lean meat. Breeding livestock are evaluated on their ability to produce progeny that will eventually yield high-quality carcasses. Therefore, muscle content and quality are important in breeding-livestock selection. Clearly, a working knowledge of carcasses and meats is key to the evaluation of all species of slaughter livestock.

MEAT QUALITY DEFINED

Meat quality has two, sometimes confusing, definitions. First, quality refers to the safety and wholesomeness of meat products. In this sense, **United States Department of Agriculture (USDA) meat inspectors** ensure quality. See figure 15-2. Inspectors observe each animal prior to slaughter, during the slaughter process, and again after slaughter to ensure the animal does not have any external or internal signs of disease. USDA inspectors are also responsible for making sure that slaughter and meat processing facilities are sanitary, thus preventing contamination of meat products after slaughter. Carcasses from animals that pass this rigorous inspection process are issued a

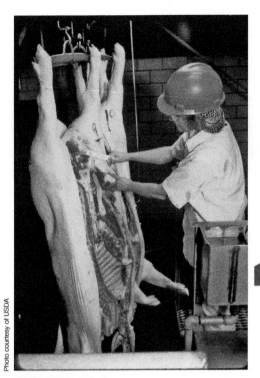

FIGURE 15-2
USDA meat inspectors ensure
the safety and wholesomeness
of meat products.

USDA inspection stamp. USDA inspection is mandatory for all meat entering the commercial food supply and is funded by the federal government.

Quality can also be defined as the predicted eating experience (taste, tenderness, flavor, juiciness) and/or composition (percent of muscle versus fat) of carcasses. **USDA meat graders** assign these predictions through quality and yield grades. While inspection is mandatory, USDA quality and yield grading are optional and must be paid for by packing plants.

BASIS FOR MEAT EVALUATION

Meat is judged using three parameters: amount of muscle present in the carcass or cut (more is better); amount of exterior fat cover (less is better); and quality of the lean meat in terms of color, texture, and marbling (intramuscular fat, more is better). The first two parameters reflect economics. A higher **lean-to-fat ratio** (percentage of lean to exterior fat) translates into more product (meat) to sell from each carcass. **Cutability** refers to percent lean from a carcass. A carcass or cut with a high lean-to-fat ratio is said to have high cutability. The third parameter, quality, attracts consumers. Consumers want bright-colored, fine-textured, well-marbled meat. Marbling, desirable to consumers, indicates flavor and juiciness of meat and helps predict a pleasurable eating experience.

A dilemma arises because, physiologically, many animals must deposit a significant or even excessive amount of exterior fat before depositing an appreciable amount of marbling. Meat cutters trim exterior fat before sale, but trimmed fat has a low economic value compared to lean meat. Ideally, meat should be highly marbled, but with a small amount of exterior fat.

CUTS OF MEAT

Whole carcasses are often judged in meats contests. However, during processing, carcasses are broken into large sections called **primal cuts**. See figures 15-3 through 15-5 for primal cut diagrams. Different primal

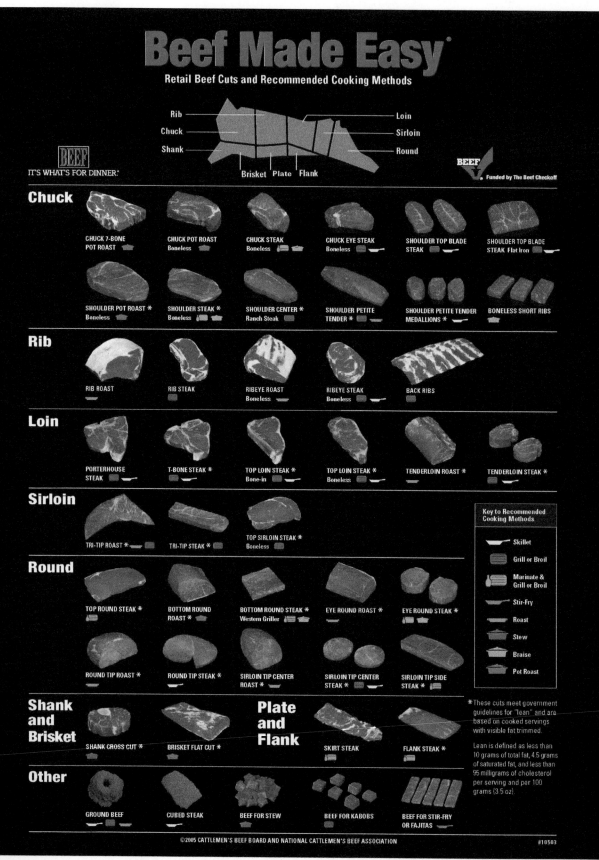

FIGURE 15-3

Cuts of beef and how to cook them.

Pork Basics

The Other White Meat

Don't be blah.®

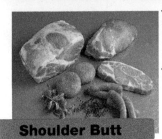

Upper row (l-r):
Bone-in Blade Roast,
Boneless Blade Roast
Lower row (l-r):
Ground Pork (The Other Burger®),
Sausage, Blade Steak

Cooking Methods
Blade Roast/Boston butt –
roast, indirect heat on grill,
braise, slow cooker
Blade Steak –
braise, broil, grill
Ground Pork –
broil, grill, roast (bake)

Shoulder Butt

Upper row (l-r):
Smoked Picnic,
Arm Picnic Roast
Lower row:
Smoked Hocks

Cooking Methods
Smoked Picnic Roast –
roast, braise
Arm Picnic Roast –
roast, braise, slow cooker
Smoked Hocks –
braise, stew

Picnic Shoulder

Top:
Spareribs
Bottom:
Slab Bacon, Sliced Bacon

Cooking Methods
Spareribs –
roast, indirect heat on
grill, braise, slow cooker
Bacon –
broil, roast (bake),
microwave

Side

Upper row (l-r):
Bone-in Fresh Ham,
Smoked Ham
Lower row (l-r):
Leg Cutlets,
Fresh Boneless Ham Roast

Cooking Methods
Fresh Leg of Pork –
roast, indirect heat on grill,
slow cooker
Smoked Ham –
roast, indirect heat on grill
Ham Steak –
broil, roast

Leg

Loin

Chops

Roasts

Upper row (l-r):
Sirloin Chop, Rib Chop, Loin Chop
Lower row (l-r):
Boneless Rib End Chop, Chef's Prime Filet™ –
Boneless Center Loin Chop, America's Cut™ –
Butterfly Chop

Cooking Methods
Cutlets (⅜ to ⅝ inch) – sauté
Thin (½ to ¾ inch thick) – grill, broil,
Thick (1¼ to 1½ inch thick) – grill, broil, roast

Upper row (l-r):
Center Rib Roast (Rack of Pork),
Bone-in Sirloin Roast
Middle:
Boneless Center Loin Roast
Lower row (l-r):
Boneless Rib End Roast,
Chef's Prime™ – Boneless Sirloin Roast

Cooking Methods
roast, indirect heat on grill, slow cooker

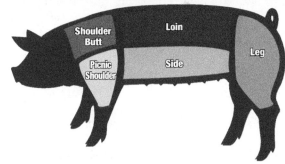

Tenderloin & Canadian-Style Bacon

Ribs

Left: Tenderloin **Right:** Canadian-Style Bacon

Cooking Methods
Tenderloin – roast, grill, pan broil
Canadian-Style bacon – roast, broil, sauté

Left: Country-Style Ribs **Right:** Back Ribs

Cooking Methods
roast, indirect heat on grill, braise, slow cooker

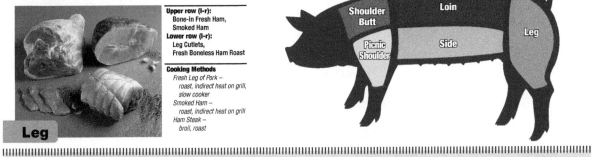

Shoulder Butt | Loin
Picnic Shoulder | Side | Leg

Roasts
No-fuss family dinner or holiday favorite

THE MANY SHAPES OF PORK

Cut Loose!

When shopping for pork,
consider cutting traditional roasts
into a variety of different shapes

Chops
Dinner, backyard
barbecue or
gourmet entree
grill, stew, braise,
broil

Cubes
Great for kabobs,
stew and chili
grill, stew, braise,
broil

Strips
Super stir fry,
fajitas and salads
grill, sauté, stir fry

Cutlets
Delicious breakfast
chops and
quick sandwiches
1/8 to 3/8 inch thick –
sauté, grill

www.TheOtherWhiteMeat.com

#03341 04/2007

Delmar/Cengage Learning

FIGURE 15-4

Cuts of pork and how to cook them.

FIGURE 15-5

Cuts of lamb and how to cook them.

cuts vary widely in their retail value. Cuts from the loin muscle are generally the most valuable because they tend to be the most tender, juicy, and flavorful. Cuts from the hindquarters rank second in value. Shoulder, neck, and cuts from the lower limbs are less valuable because they generally contain more connective tissue and are less tender. When whole carcasses are evaluated, more emphasis should be placed on the primal cuts of the highest value.

Primal cuts themselves may also be judged in a meats judging contest. In high-priced cuts (such as pork loins, and beef or lamb ribs), muscle quality must be emphasized, but not necessarily at the expense of cutability. Primal cuts of intermediate price, such as beef rounds, pork legs/hams or lamb legs, should be judged on cutability, with muscle quality taking secondary importance. Lower-priced primals, such as pork shoulders or beef chucks, should be judged with a heavy emphasis on cutability.

Primal cuts are often further broken into **subprimal cuts** before being packaged, boxed, and shipped to restaurants, supermarkets, or institutions such as schools and hospitals. Upon arrival at their destination, subprimal cuts are further reduced into retail cuts before cooking and consumption. By looking at a retail cut of meat, contest participants may be asked to identify the species, primal cut, subprimal cut, and the best way to cook that piece of meat. Figures 15-3 through 15-5 provide all that information.

Institutional Meat Purchasing Specifications (IMPS) have been adopted by the meat industry to standardize various primal and subprimal cuts. Standardization became necessary because of increased boxed meat sales. For instance, a restaurant that specializes in prime rib does not need a whole beef carcass, so only beef ribs, which come packaged one or two to a box, are purchased. IMPS were adopted to ensure consistency of boxed meat. IMPS cuts are numbered in a series.

100 series - fresh beef cuts

200 series - fresh lamb and mutton

300 series - fresh veal and calf

400 series - fresh pork

500 series - cured, smoked, or cooked pork

600 series - cured, dried, or smoked beef

700 series - edible by-products

800 series - sausage by-products

1,000 series - portion-cut meat products.

Numbers within the series identify specific cuts. Practically, this numbering system ensures that a box of 107 beef ribs ordered by a restaurant this week contains exactly the same meat cut, trimmed the same way, as the 107 beef ribs the restaurant ordered last week.

Individual retail cuts are fair game in a judging contest. A class of four retail cuts may be judged using the same three criteria—muscling, trimness, and quality—used to judge carcasses or primal cuts. Refer to figures 15-3 through 15-5 for a breakdown of beef, pork, and lamb carcasses into retail cuts.

COOKING DIFFERENT CUTS

Judging contestants are sometimes asked the best way to cook a certain piece of meat. Generally, a steak or thinly sliced pieces of meat are broiled, pan broiled, or pan fried at high temperature for short duration using dry cooking methods. Steaks are naturally juicy and tender. Short-duration, high-heat cooking seals in the juiciness and retains the inherent tenderness.

Roasts or large, tender, pieces of meat should usually be roasted. Roasting involves a moist, long-duration, lower-heat cooking method. During the roasting process, the roast constantly self-bastes, retaining the meat's original moisture content.

Less tender pieces of meat can be braised. Braising involves seasoning and browning of the meat, followed by simmering in a covered, liquid-filled container. This helps break down the connective tissue, to make the meat more tender.

BEEF JUDGING

Contestants in meat judging contests assign both yield and quality grades to beef carcasses using the same criteria as USDA graders. Through **yield grades** (USDA estimate of cutability) and **quality grades** (USDA estimate of eating experience based mostly on marbling and age), the value of an entire beef carcass can be predicted. Contestants in a meat judging contest are asked to yield and quality grade a rail (15 carcasses) of cattle for a score. The basis for both grades relies primarily on an appraisal of the ribeye muscle cut between the 12th and 13th ribs.

YIELD GRADING

Yield grades are based on a score of one to five with one being an extremely lean, heavily muscled carcass, and five being an extremely fat, lightly muscled carcass. Most slaughter cattle have yield grades between two and four. The exact formula for calculating yield grade is:

Yield grade = 2.5 + (2.5 × adjusted 12th rib fat thickness, inches) + (0.0038 × hot carcass weight, pounds) + (0.2 × kidney, pelvic, and heart fat, percent) − (0.32 × 12th rib ribeye area, square inches)

However, most contestants are instructed in the following method for yield grading a beef carcass quickly.

STEP 1

The amount of fat covering the ribeye gives a good approximation of the amount of fat on the entire carcass. Therefore, a measurement of the exterior fat cover, 3/4 the distance over the ribeye at the 12th rib, is used to calculate a **preliminary yield grade (PYG)**. See figure 15-6. If this measurement is drastically more or less than the fat cover over the lower rib, the PYG may be adjusted up or down. PYG is the base yield grade that will be adjusted during steps 2 and 3. Meats judging contestants must be able to accurately estimate the fat depth over the ribeye without using a measuring tool.

Fat thickness	PYG
0	2.0
0.2	2.5
0.4	3.0
0.6	3.5
0.8	4.0
1.0	4.5

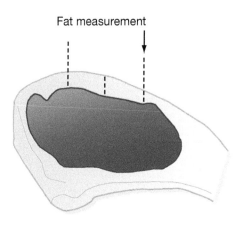

Fat measurement

FIGURE 15-6

Fat thickness is measured 3/4 the distance over the ribeye muscle and the measurement converted to preliminary yield grade (PYG) in beef carcasses.

Delmar/Cengage Learning

Animal Science Facts

A plastic or cellophane grid can be used to measure ribeye or loineye area in a carcass. Simply lay the grid on the cut surface of the muscle and count the number of dots that fall inside the muscle area. Divide the number of dots by 10 (for beef grids) or 20 (for lamb and pork grids) to calculate area in square inches. A more accurate method to determine the area is with a device called a compensating polar planimeter. This device is used to outline a tracing of the eye muscle. The area (in square inches) can be calculated by subtracting the initial reading from the final reading.

STEP 2

The PYG is adjusted for the size of the ribeye compared with carcass weight. If the ribeye is bigger than expected for the weight of the carcass, the PYG is adjusted down (remember, a lower numerical yield grade is better). If the ribeye is smaller than expected, the PYG is adjusted up. See figure 15-7 for expected ribeye areas. Again, contestants are expected to be able to estimate ribeye area in square inches without using a measuring tool. Many successful contestants measure the number of square inches in the palm of their hand, then use that known area to help estimate the number of square inches of ribeye in a carcass.

PYG is adjusted by 0.3 for every square inch of ribeye area above or below the expected target (or 0.1 for every 0.3 square inch of ribeye area). For example, a 600-pound carcass with a 12.0 square inch ribeye would result in a subtraction of 0.3 from the PYG. A 600-pound carcass with a 10.0 square inch ribeye would result in the addition of 0.3 to the PYG. Remember, more muscle means subtract from the PYG.

Carcass Wt.	Expected Ribeye (in.^{2})
500	9.8
550	10.4
600	11.0
650	11.6
700	12.2
750	12.8
(Remember, 600 lb = 11.0 in.^{2} and add or subtract 0.3 in.^{2} for every 25 lb of carcass weight)	

Delmar/Cengage Learning

FIGURE 15-7

Expected ribeye areas for various carcass weights.

STEP 3

Finally the PYG may be adjusted for the amount of internal fat present in the carcass. Internal fat in beef carcasses is more often called **kidney, pelvic, and heart fat (KPH)**. KPH of 3.5 percent of carcass weight is considered average. If the carcass contains more than 3.5 percent KPH, the PYG is adjusted up. If the carcass contains less than 3.5 percent KPH, the PYG is adjusted down. For example 3.5 percent of a 600-pound carcass would be 21 pounds. Contestants attempt to estimate the number of pounds of internal fat in a carcass and compare the estimate to the number of pounds equaling 3.5 percent of the carcass weight. See figure 15-8 for adjustments.

Example Yield Grade Calculation:

ACTUAL MEASUREMENT OR ESTIMATE		
Carcass weight	700	pounds
Adjusted fat thickness	0.40	inches
Ribeye area	11.6	square inches
KPH%	2.5	percent

Calculation:

1. PYG (from table figure 15-6) = 3.0
2. Ribeye adjustment
   expected ribeye (from figure 15-7) = 12.2
   actual ribeye = 11.6
   Adjustment $(12.2 - 11.6) \times 0.3 = +0.2$
3. KPH adjustment (from figure 15-8) = -0.2
4. Final yield grade = 3.0 (PYG) + 0.2 (Ribeye adjustment) − 0.2 (KPH adjustment) = 3.0

Note—When calculating ribeye adjustments, round all numbers to the nearest 0.1.

Obviously, calculating yield grades takes practice. In a contest, participants must estimate fat, ribeye and KPH measurements, and then calculate a yield grade based on their own estimates. Carcass weights will be provided.

QUALITY GRADING

Beef quality grades are primarily based on two factors: the physiological age of the carcass and the amount of marbling present in the cut surface of the ribeye.

KPH%	Adjustment to PYG
1.0	−0.5
1.5	−0.4
2.0	−0.3
2.5	−0.2
3.0	−0.1
3.5	0
4.0	+0.1
4.5	+0.2

Delmar/Cengage Learning

FIGURE 15-8

Beef yield grade adjustments for varying percentages of kidney, pelvic, and heart fat (KPH).

The first factor in quality grading is physiological maturity. As any animal gets older, muscle tenderness decreases because the amount of tough connective tissue increases. Because actual animal age is rarely known, quality graders attempt to estimate the age of a carcass by looking for clues hidden in the bones.

Buttons are pieces of cartilage along the split surface of the carcass over the rib. See figure 15-9. Buttons become harder and more calcified as cattle age. Along with buttons, the appearance of the ribs inside the carcass changes with age. As cattle get older, ribs that were once round and red, become flat and white. Also, as cattle age, the lumbar vertebrae in the sirloin region fuse. See figure 15-10 for locations of bones used in aging carcasses.

Bone clues are used to give each carcass a maturity score. Maturity scores range from A (youngest) to E (oldest). Each score is further divided into

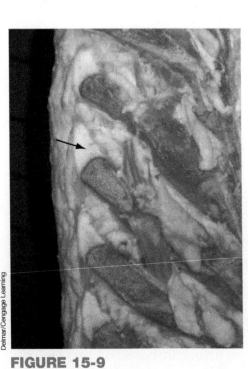

FIGURE 15-9

White pieces of cartilage perched at the end of the bony spinal processes are called buttons.

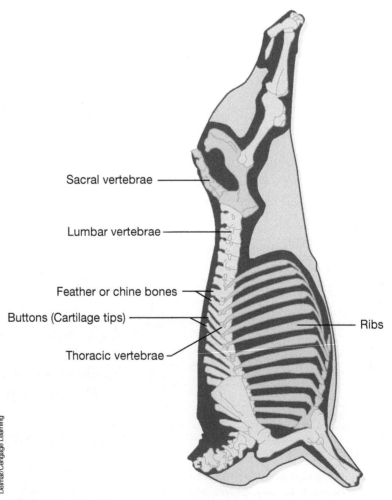

Sacral vertebrae

Lumbar vertebrae

Feather or chine bones

Buttons (Cartilage tips)

Thoracic vertebrae

Ribs

FIGURE 15-10

Bones used to estimate maturity in beef carcasses.

thirds. The youngest third of the A maturity would be classified A–, the middle third of the A maturity would be A0, and the oldest third, A+. All other maturity levels are divided similarly. See figure 15-11 for comparison of bone clues with maturity scores. Color and texture of lean are used as tiebreakers if a carcass is on the borderline between two maturities. Carcasses with bright, finely textured lean receive the younger score. Carcasses with dark, coarsely textured lean receive the older score.

Marbling is a subjective measurement based upon the amount of intramuscular fat visible on the cut surface of the ribeye. The amount of marbling depends on feeding programs and genetics. In general, carcasses with higher degrees of marbling are more valuable because increased amounts of marbling predict a more enjoyable eating experience. Marbling is divided into nine categories (listed from the most marbled to the least): abundant, moderately abundant, slightly

USDA MATURITY GROUP	SACRAL VERTEBRAE	LUMBAR VERTEBRAE	THORACIC[1] VERTEBRAE	RIBS	AGE[2]
A–	Distinct separation	No ossification	No ossification	Slight tendency toward flatness	9 mos.
A+/B–	Completely fused	Nearly completely ossified	Cartilages show some evidence of ossification (5% ossified)	Slightly wide and slightly flat	30 mos.
B+/C–	Completely fused	Completely ossified	Cartilages are partially ossified (20 to 30% ossified)		42 mos.
C+/D–	Completely fused	Completely ossified	Outlines of cartilages are plainly visible (65 to 75% ossified)	Moderately wide and flat	72 mos.
D+/E–	Completely fused	Completely ossified	Outlines of cartilages are barely visible (95% ossified)	Wide and flat	96 mos.

[1] Descriptions refer to the uppermost three thoracic vertebrae in the forequarter (in the region of the 10th, 11th, and 12th ribs.

[2] Approximate chronological age equivalent.

Delmar/Cengage Learning

FIGURE 15-11

Comparison of bone clues to maturity with USDA maturity group.

abundant, moderate, modest, small, slight, traces, practically devoid. See figure 15-12 for some of the more common marbling degrees.

Each marbling category further subdivides into percentages from 0 to 100 percent in 10 percent

Slight

Small

Modest

Moderate

Slightly Abundant

Moderately Abundant

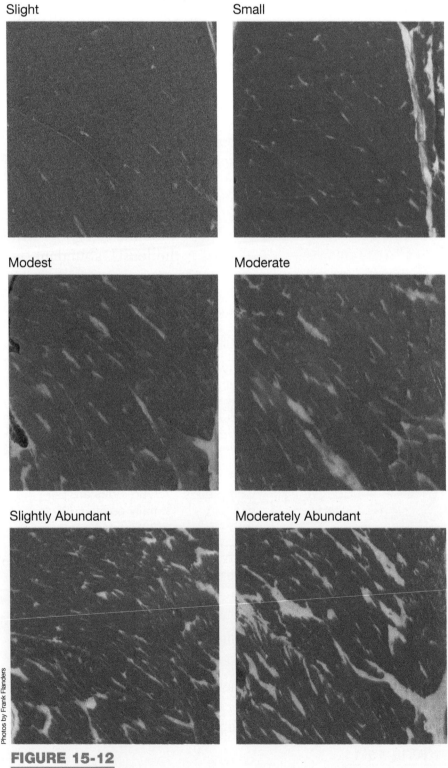

Photos by Frank Flanders

FIGURE 15-12
USDA degrees of marbling.

increments. For example, small marbling could be designated small10, small20, small30, and so on. Small30 is a higher marbling score than small10.

Marbling score combined with maturity score determine the final quality grade. The USDA quality grades (from most to least desirable) are Prime, Choice, Select, Standard, Commercial, Cutter, and Canner. See figure 15-13. The Prime, Choice, and Standard grades can be applied only to cattle with A or B maturity scores. The Select quality grade can be applied only to carcasses with an A maturity score.

Some quality grades are further broken down into halves or thirds of a grade. For example the Choice grade is divided into thirds: For A maturity, low choice aligns with small marbling, average choice aligns with modest marbling, and high choice aligns with moderate marbling. The select grade is divided into high and low select for slight marbling. Slight 0 to slight 50 aligns with low select while the upper portions of the slight marbling degree cast a carcass into the high select grade. The prime grade is also divided

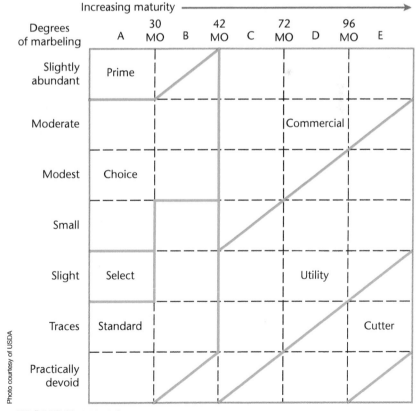

Photo courtesy of USDA

FIGURE 15-13

This table combines marbling with maturity score to arrive at the final USDA beef quality grade.

into thirds, but carcasses grading higher than low prime (slightly abundant marbling) are rare. Refer to figure 15-13 and compare the following carcasses. Line up marbling score on the left column with maturity score along the top. Trace both lines into the chart until they intersect. Use figure 15-13 to determine quality grade from the several examples listed in the following table.

MARBLING SCORE	MATURITY SCORE	QUALITY GRADE
Slightly Abundant10	A0	Prime−
Slightly Abundant10	B+	Choice +
Moderate10	A0	Choice +
Small50	A0	Choice −
Slight80	A0	Select +
Slight80	B−	Standard

Successful meat graders memorize figure 15-13 and can accurately quality grade cattle without visual aids.

JUDGING BEEF CARCASSES

Once yield and quality grading are mastered, judging beef carcasses becomes very easy. Simply yield and quality grade the carcasses, and then rank them on a combination of yield and quality. The carcass with the best combination of yield and quality grade would be placed first, the second best combination, second, and so on. See figure 15-14 for parts of a beef carcass.

BEEF TERMINOLOGY AND REASONS

Terminology for beef carcass judging relies on a certain comfort with the descriptive names of carcass parts. For instance, if a certain carcass is fatter over the sirloin, the person delivering the reasons has to know the sirloin location. Study and memorize the parts presented in figure 15-14.

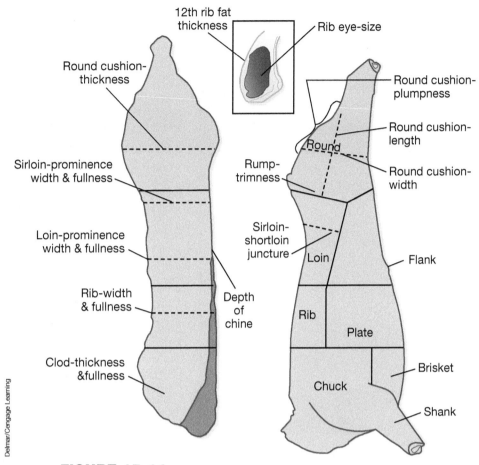

Delmar/Cengage Learning

FIGURE 15-14

Parts of a beef carcass.

Meat reasons format reasons differs little from the generic reasons used in Chapter 13. Opening statements for each pair should reveal the most important reasons for the placing. Broad opening statements should be followed by specific examples that reinforce the opening statement. The three key areas that must be covered in each pair are differences in muscling, trimness, and quality. Comparative terms should be used within the discussion of each pair. Grants should be given when necessary.

Whole Beef Carcasses

Use figure 15-14 to assist in the following section. The following are places to look for muscling differences on beef carcasses and some descriptive terms to use in a set of reasons:

Chuck—heavier muscled, meatier, thicker

Ribeye—larger, fuller, more oval-shaped

Sirloin—wider

Rump—wider, fuller

Round—deeper, wider, or thicker cushion

The following are places to look for trimness differences on beef carcasses, and some descriptive terms to use in a set of reasons:

Ribeye fat—less fat at the 12th rib

KPH fat—lower percent KPH, less kidney, pelvic, and heart fat

Brisket, chuck, lower rib, flank, loin edge, sirloin-shortloin junction, sirloin, rump, round—trimmer

Less fat over clod, less cod or udder fat

The following are places to look for quality differences on beef carcasses, and some descriptive terms to use in a set of reasons:

Roundness of ribs—more youthful bone

Buttons—showing less ossification

Color of fat—whiter, flakier

Ribeye—higher degree of marbling, more evenly distributed marbling, brighter cherry red color, more youthful color, finer-textured lean.

Beef Primal Cuts

Chucks, ribs, loins, or rounds can be judged using the principles learned from yield and quality grading. Chucks and rounds are judged with a heavy emphasis on yield or cutability. Classes of chucks and rounds should be placed almost entirely on cutability with the heaviest muscled, leanest cuts placed at the top of the class. Loins and ribs however, are judged with an emphasis on quality. Marbling, color, and texture of lean are more important than yield on these more expensive primal cuts. However, even highly marbled loins and ribs from a yield grade 4 or 5 carcass may be discounted depending on current beef pricing structure. Thus contestants should be aware of the current **choice/select spread** (difference in carcass price between choice and select carcasses) prior to competing in a meat judging contest.

Chucks

Use figure 15-15 to assist with the following text. The following are places to look for muscling differences

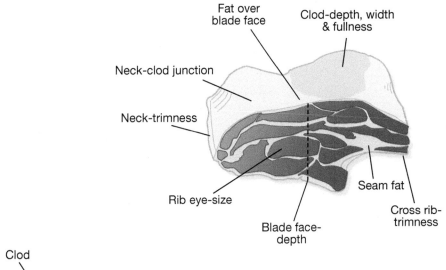

BLADE END

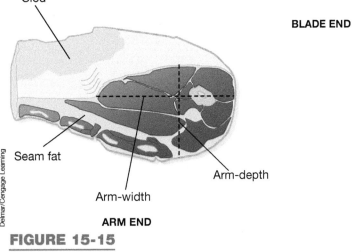

ARM END

FIGURE 15-15

Parts of a beef chuck.

on beef chucks and some descriptive terms to use in a set of reasons:

> Blade end or face—bigger blade eye, wider or deeper blade face
>
> Clod—fuller
>
> Neck—fuller, thicker
>
> Arm end or face—wider, deeper

The following are places to look for trimness differences on beef chucks and some descriptive terms to use in a set of reasons:

> Blade end or face—less fat over blade end, less seam fat in the blade face
>
> Neck-clod junction—trimmer
>
> Arm end or face—less fat over arm end, less seam fat in the arm face

The following are places to look for quality differences on beef chucks and some descriptive terms to use in a set of reasons:

Blade end or face—brighter-colored lean, higher degree of marbling

Arm end or face—brighter-colored lean

Fat color—whiter, flakier

Ribs

Use figure 15-16 to assist with the text presented here. The following are places to look for muscling differences on beef ribs and some descriptive terms to use in a set of reasons:

Ribeye—larger, fuller, more oval-shaped, longer, wider

Chine—deeper

Blade end or face—wider, deeper, more exposed lean, bigger blade eye

The following are places to look for trimness differences on beef ribs and some descriptive terms to use in a set of reasons:

Loin end, ribeye, lower rib, back, lower blade—trimmer

Blade end—trimmer, less seam fat

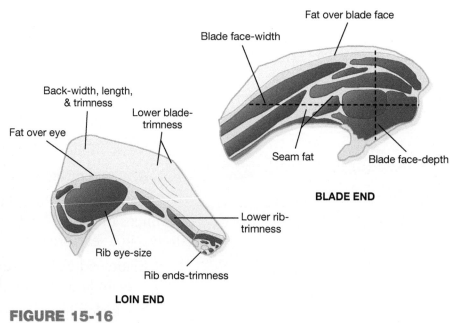

FIGURE 15-16

Parts of a beef rib.

The following are places to look for quality differences on beef ribs and some descriptive terms to use in a set of reasons:

Ribeye—same as for beef carcasses

Loins

Use figure 15-17 to assist with beef loin judging. The following are places to look for muscling differences on beef loins and some descriptive terms to use in a set of reasons:

Loineye—wider, deeper, larger

Back—wider, longer

Chine—deeper

Sirloin—fuller

Sirloin end—more exposed lean in top or bottom sirloin

The following are places to look for trimness differences on beef loins and some descriptive terms to use in a set of reasons:

Loineye, loin edge, flank edge, sirloin-shortloin junction—trimmer

Sirloin end—trimmer, less seam fat

Pelvic area—less pelvic or channel fat

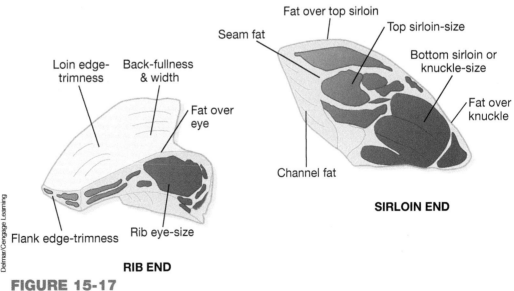

Delmar/Cengage Learning

FIGURE 15-17

Parts of a beef loin.

The following are places to look for quality differences on beef loins and some descriptive terms to use in a set of reasons:

Loineye—same as for beef carcasses

Sirloin end—same as for beef carcasses

Rounds

Use figure 15-18 to assist with judging beef rounds. The following are places to look for muscling differences on beef rounds and some descriptive terms to use in a set of reasons:

Sirloin end or face—more lean area exposed, deeper, wider

Cushion—deeper, wider, longer

Heel—meatier, fuller, plumper

Shank—shorter (indicates more muscle)

The following are places to look for trimness differences on beef rounds and some descriptive terms to use in a set of reasons:

Sirloin end or face, rump, knuckle, cushion, heel—trimmer

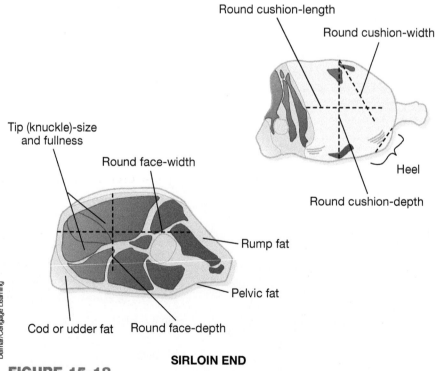

Delmar/Cengage Learning

FIGURE 15-18

Parts of a beef round.

Sirloin end—less seam fat

Cod or udder, pelvic fat—less

The following are places to look for quality differences on beef rounds and some descriptive terms to use in a set of reasons:

Sirloin end or face—marbling, firmness, texture, color (same as beef carcasses)

General Introductory Beef Carcass or Cut Terms

Higher percentage boneless, closely trimmed retail cuts

Higher cutability

Higher percentage trimmed steaks and roasts

More total pounds of trimmed steaks and roasts

Higher percentage of higher-priced cuts

Higher yielding

Have a more desirable yield grade

Higher, more desirable quality grade

SAMPLE BEEF CARCASS REASONS

Notes for all meat judging should be recorded for the three main criteria (muscling, trimness, and quality) for each pair. In a given pair, only the criteria that were important to placing the class should be discussed. Also, meats reasons may be written rather than given orally, so good penmanship and grammar are essential.

Beef Carcasses Placing: 1–2–3–4

FIRST 1/2—In this class of beef carcasses, 1 placed over 2 as 1 more ideally combined cutability and quality. 1 was heavier muscled as demonstrated by a thicker, fuller cushioned round, which carried into a deeper heel. In addition, 1 was a higher-quality carcass as shown by a more youthful, cherry red, finer-textured lean. 1 had an additional advantage in trimness particularly over the ribeye, lower rib, loin edge, and chuck. Thus, 1 would yield a higher percentage of closely trimmed, consumer acceptable cuts. I realize 2 had a clear advantage in the degree of marbling and was trimmer internally, displaying less kidney and pelvic fat.

SECOND 2/3—2 easily placed over 3, as 2 was a significantly higher-quality, heavier-muscled carcass that would yield meatier, trimmer retail cuts with more consumer appeal. 2 displayed a much higher degree of more evenly distributed marbling. In addition, 2 was more youthful in its appearance as evidenced by rounder, redder ribs and softer, whiter buttons. 2 also featured a larger, more oval-shaped ribeye, a deeper cushioned round, and a thicker clodded chuck. I concede that 3 was decidedly trimmer over the ribeye, lower rib, rump, and round. 3 also had an advantage in cod, kidney, and heart fat.

THIRD 3/4—I preferred 3 over 4 as 3 would yield a significantly higher percentage of high-value, trimmed primals. 3 displayed less external fat over the round, rump, loin edge, and was especially trimmer over the ribeye and lower rib. In addition, 3 was trimmer over the chuck and displayed less cod and kidney fat. I admit 3 had a larger ribeye and a thicker, clodded chuck, partially due to fat.

FOURTH 4—I admit 4 was acceptable in terms of cutability. However, it displayed the narrowest, lightest muscled round in the class, which, coupled with the lowest quality in the class, would make it the least desirable to the profit-minded retailer.

PORK JUDGING

Judging of pork carcasses and cuts relies, to a very high degree, on cutability. While USDA grades for pork carcasses exist, they are not used in the industry. Hence, there is no standard method for yield and quality grading pork. Pork packing plants use a variety of company-specific formulas for assessing carcass value but the principles discussed in the beef judging section apply. Pork carcasses or cuts with high cutability and quality are preferred over fat, low-quality carcasses and cuts. See figure 15-19.

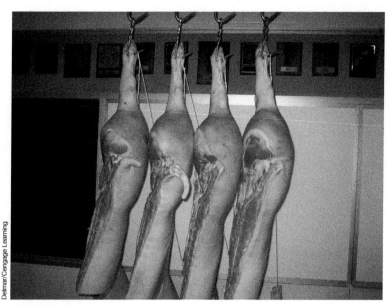

Delmar/Cengage Learning

FIGURE 15-19

Class of four pork carcasses that might be used in a meats judging contest.

JUDGING PORK CARCASSES

Hams and loins are traditionally the most highly priced pork cuts. Therefore, carcasses with the highest percentage of lean in the ham and loin with acceptable quality are preferred. Carcasses are split and may be judged ribbed (loin muscle cut and displayed at the 10th rib) or not ribbed. When carcasses are ribbed, fat thickness should be evaluated 3/4 the distance over the ribeye muscle, similar to beef carcasses. When carcasses are not ribbed, judges must make estimates of the size of the loineye and total carcass muscle based on width and thickness of the loin and ham, as well as the amount of lumbar lean, depth of chine, or overall carcass thickness. See figure 15-20 for parts of a pork carcass.

Trimness is evaluated primarily by the amount of backfat present, either at the 10th rib or as average back fat over the entire split surface of the carcass. Other indications of carcass trimness include fat over the ham collar, the ham-loin junction, elbow pocket, jowl, sternum, and belly edge.

Quality can be easily seen in ribbed carcasses by evaluating the color (grayish pink is ideal), firmness, and marbling of the exposed loin muscle. In unribbed carcasses, color and firmness can be evaluated in the

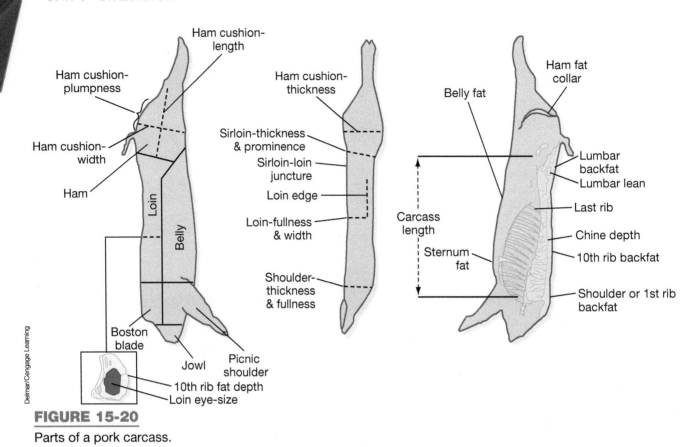

Delmar/Cengage Learning

FIGURE 15-20

Parts of a pork carcass.

lumbar lean, over the ham face, and between the ribs where feathering (fat streaking) is an indication of quality.

Pork Terminology and Reasons

Again, success in giving pork reasons relies on the proper use of terminology. Study and memorize the parts of a pork carcass, as well as the parts of a ham.

Whole Pork Carcasses

Use figure 15-20 to assist with the following section. The following are places to look for muscling differences in pork carcasses and some descriptive terms to use in a set of reasons:

> Ham—fuller, meatier, thicker, more bulging, wider, plumper
>
> Sirloin—fuller, plumper, meatier
>
> Lumbar lean—greater exposed area, larger area
>
> Loin—fuller, wider, deeper chined
>
> Loineye (if ribbed)—larger, shapelier
>
> Shoulder—meatier, thicker fleshed, thicker

The following are places to look for trimness differences in pork carcasses and some descriptive terms to use in a set of reasons:

Backfat—trimmer at the lumbar, last rib, and first rib (if not ribbed), trimmer over the loineye (if ribbed)

Ham collar, flank, belly, sternum, jowl—trimmer

Elbow pocket—more definition

Ham–loin junction—more definition

The following are places to look for quality differences in pork carcasses and some descriptive terms to use in a set of reasons:

Loineye (if ribbed)—finer-textured, more evenly dispersed marbling, higher degree of marbling, firmer-textured lean, more ideal grayish-pink color

Lumbar lean—higher degree of marbling

Belly—fuller, firmer, thicker (these are important quality characteristics for bacon production)

Rib feathering—higher degree, greater quantity, more extensive

Pork Primal Cuts

Fresh hams are the most commonly judged primal cuts of pork. As with carcasses, the ham that best combines muscling and trimness with acceptable quality wins the class. See figure 15-21 for the parts of a fresh ham.

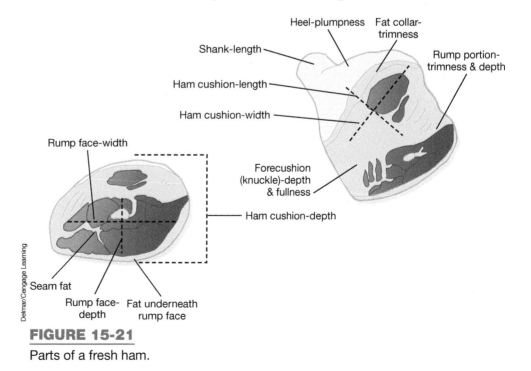

FIGURE 15-21

Parts of a fresh ham.

Hams

Use figure 15-21 to assist with the following section. The following are places to look for muscling differences in fresh hams and some descriptive terms to use in a set of reasons:

Shank—shorter (indicates more muscle)

Rump or butt face—deeper, wider, greater amount of exposed lean

Forecushion or knuckle—fuller, meatier, plumper

Cushion or center section—longer, wider, deeper, plumper, more bulging

The following are places to look for trimness differences in fresh hams and some descriptive terms to use in a set of reasons:

Under the butt face, along the rump face, over the collar, over the cushion—trimmer

Seam fat—less visible

The following are places to look for quality differences in fresh hams and some descriptive terms to use in a set of reasons:

Marbling in the rump face—more, a higher degree, more visible

Texture of exposed lean—firmer, finer-textured lean

Color of exposed lean—more uniform grayish-pink

General Introductory Pork Terms
Carcasses

A higher-yielding carcass

Carcass that would yield a higher percent muscle

Carcass that would yield a higher percent ham and loin

Carcass with a higher lean to fat ratio

Carcass yielding a higher percentage of the trimmed four lean cuts

Hams

Ham that would yield a higher percentage of muscle

Ham that would yield a higher percentage of center cut slices

Ham with a higher lean to fat ratio

Sample Pork Carcass Reasons

Pork Carcasses Placing: 1–2–3–4

FIRST 1/2—I placed 1 over 2 as 1 combined muscling and trimness to a higher degree. 1 was thicker through the center portions of the ham and significantly wider through the sirloin. In addition, 1 was leaner at the last rib and displayed less sternum fat. I admit that 2 was trimmer over the ham collar and exhibited more rib feathering.

SECOND 2/3—2 placed over 3 as 2 was simply a higher-cutability carcass with a higher lean to fat ratio. 2 possessed less fat at the first rib, last rib, and last lumbar and was particularly leaner over the loin edge and at the ham collar. Furthermore, 2 displayed a shorter-shanked ham and more exposed lumbar lean. I concede 3 did exhibit more marbling in the lumbar lean.

THIRD 3/4—3 more completely combined muscling and quality to a higher degree than 4. 3 displayed a decidedly thicker-cushioned ham, a thicker sirloin, and a more prominent shoulder. Furthermore, 3 displayed a more desirable grayish-pink color in the lumbar lean, as well as more rib feathering. I admit 4 was trimmer at the jowl and over the ham collar.

FOURTH 4—I placed 4 last as it was the lightest muscled, poorest quality carcass in the class. 4 was narrowest in the shoulder, loin, and ham. Also, 4 showed the poorest colored exposed lean and would be the least appealing to the quality-conscious consumer.

LAMB JUDGING

USDA provides a mechanism for yield and quality grading lamb carcasses, similar to that of beef carcasses. However, as with pork, the judging of lamb carcasses and cuts relies to a very high degree on cutability.

JUDGING LAMB CARCASSES

Some lamb carcasses are ribbed between the 12th and 13th ribs, as are beef carcasses, exposing the ribeye, however, lamb carcasses are not split vertically. Others are not ribbed and displayed whole. When carcasses are ribbed, fat estimates or measurements should be made 1/2 the distance over the ribeye.

Lambs are yield graded using the 12th rib fat measurement. Simply insert the 12th rib fat depth into the formula:

(Fat depth × 10) + 0.4 = yield grade

Other areas to check for external fat deposit include the shoulder, sirloin, dock, leg, cod or udder, flank, or breast. The expensive cuts of lamb are from the hindsaddle, (portion of the carcass to the rear of the 12th rib). Therefore, differences in muscling of the hindsaddle should be evaluated with care. Muscling differences are most easily seen in the ribeye (ribbed carcasses), dimensions of the leg, and width of the loin, sirloin and rump (unribbed carcasses). See figure 15-22.

The color of lamb should be bright reddish pink. Other parameters indicating quality include flank streaking and rib feathering.

FIGURE 15-22

Class of four lamb carcasses that might be used in a meats judging contest.

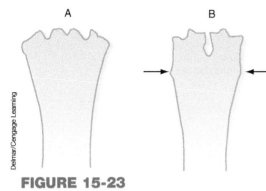

FIGURE 15-23

Comparison of a break joint (A) with a spool joint (B).

Lambs are also quality graded using a combination of overall conformation and flank streaking. A vast majority of the lambs that are graded reach choice or prime. However, an age parameter is in place so that only young lambs can be graded.

As lambs mature to one year of age, the growth plates of the long bones begin to calcify and harden, signaling the end of skeletal growth. When the front legs of a young, immature lamb are removed, the resulting red, porous bone left showing is called a **break joint**. The break joint is actually the joint separation at the growth plate rather than the true joint. As lambs approach 12 months of age, the break joints ossify and, at some point, will no longer separate. See figure 15-23. After the ossification process is complete, the front leg removal occurs at the true joints, called **spool joints**. Spool joints are bone white in color.

LAMB TERMINOLOGY AND REASONS

Lambs have a few different parts as compared to those studied previously in beef and pork. Study and memorize the parts in figure 15-24.

Lamb Carcasses

Use figure 15-24 to assist with the following section. The following are places to look for muscling differences in lamb carcasses and some descriptive terms to use in a set of reasons:

> Leg—fuller, meatier, thicker, more bulging, wider, plumper
>
> Sirloin—fuller, plumper, more prominent
>
> Rack or rib—fuller, wider
>
> Rib eye (if ribbed)—larger, shapelier
>
> Shoulder—fuller, thicker

The following are places to look for trimness differences in lamb carcasses and some descriptive terms to use in a set of reasons:

> Ribeye, lower rib, leg cushion, sirloin, dock, loin edge, rack, shoulder, internal, cod or udder, cod, breast, flank - trimmer, displayed less fat
>
> Breast, flank—trimmer, displayed less fat

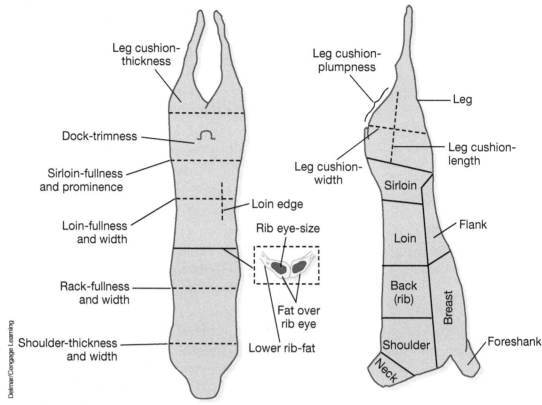

Delmar/Cengage Learning

FIGURE 15-24

Parts of a lamb carcass.

The following are places to look for quality differences in lamb carcasses and some descriptive terms to use in a set of reasons:

Ribeye (if ribbed)—more youthful, brighter-colored lean, firmer-textured lean, higher degree of marbling

Rib feathering—higher degree, greater quantity, more extensive

Flank streaking (lacing)—greater primary and secondary flank streaking, more finely dispersed.

Lamb Primal Cuts

Primal cuts of lamb are judged sparingly. Terms are similar to those used for beef primal cuts (but substituting leg for round, etc.)

General Introductory Lamb Terms

Carcass that would yield a higher percentage of trimmed leg and loin

Carcass that would yield the highest percentage boneless, trimmed retail cuts

Carcass that would yield a higher percent hindsaddle

Sample Lamb Carcass Reasons

Lamb Carcasses Placing: 1–2–3–4

FIRST 1/2—1 placed over 2 as 1 was a significantly heavier-muscled carcass that would yield a higher percent hindsaddle. 1 displayed more inside and outside flare in a heavier-muscled leg, a wider sirloin, and a decidedly larger ribeye. In addition, 1 displayed a more youthful appearing ribeye and possessed more primary and secondary flank streaking. I admit 2 was somewhat trimmer over the leg, dock, cod, and breast.

SECOND 2/3—2 placed over 3 in a close decision as 2 was a trimmer, higher-cutability carcass. 2 exhibited less fat at the ribeye, over the dock, leg, and shoulder. Furthermore, 2 had a longer cushioned leg. I concede 3 possessed a larger ribeye, and was a higher-quality carcass, as evidenced by a higher degree of ribeye marbling and more rib feathering.

THIRD 3/4—3 handily placed over 4 as 3 more ideally combined trimness and muscling into a higher-cutability carcass. 3 was trimmer at the ribeye, over the sirloin, and over the leg. Furthermore, 3 had decidedly less pelvic fat. Moreover, 3 was meatier at the ribeye, over the sirloin, and possessed a wider, thicker cushioned leg. I admit 4 showed more primary flank lacing and a higher degree of marbling at the eye.

FOURTH 4—I realize 4 showed the brightest-colored lean and the highest degree of marbling; however, 4 simply combined trimness and muscling to the lowest degree. 4 exhibited the most fat over the ribeye, loin, dock, leg, and shoulder. Also, 4 displayed the smallest ribeye in the class. Accordingly, 4 would yield the lowest percentage of closely trimmed retail cuts.

SUMMARY

Carcasses are broken down into primal, subprimal, and retail cuts. IMPS cuts are standardized primal and subprimal cuts used in the institutional and restaurant trade. Tender cuts of meat should be cooked with high, dry heat for a short duration. Tougher cuts require a longer period of moist heat for optimum flavor and tenderness. Yield grades are used to differentiate between lean and fat carcasses. Quality grades attempt to predict the flavor and tenderness of meat. Age is an important consideration when quality grading beef and lamb. Meat is judged on a combination of muscle quantity, leanness, and muscle quality. Knowledge of comparative terminology of carcasses and cuts is very useful to meat judges.

CHAPTER REVIEW

EXPERIENTIAL LEARNING OPPORTUNITIES

1. Students interested in obtaining experience in meat judging may consider working at a retail meat counter for a supervised agricultural experience program.

2. Students carrying market livestock projects could enter an animal in a carcass competition.

DEFINE ANY TEN KEY TERMS

break joint

buttons

choice/select spread

cutability

Institutional Meat Purchasing
  Specifications (IMPS)

kidney, pelvic, and heart fat (KPH)

lean-to-fat ratio

preliminary yield grade (PYG)

primal cuts

quality grades

spool joint

subprimal cuts

USDA meat graders

USDA meat inspectors

yield grades

QUESTIONS AND PROBLEMS FOR DISCUSSION

1. Three primary factors influence the judging of meat. Name them.

2. List the primal cuts of lamb.

3. Primal cuts taken from the _____ are the most valuable.

4. Write the IMPS series number for fresh beef.

5. Steaks should be cooked using:

   a. High heat

   b. Medium heat

   c. Low heat

6. Braising:

    a. involves seasoning and browning of meat

    b. involves simmering in liquid

    c. helps tenderize meat

    d. all of the above

7. Differentiate between methods for cooking a steak and a roast.

8. Yield grade indicates _____.

9. Beef quality grade is based on _____ and _____.

10. _____ percent internal fat is considered average for beef carcasses.

11. Name the nine marbling categories.

12. Maturity scores in beef carcasses range from _____ (youngest) to _____ (oldest).

13. The highest quality grade is _____.

14. Trimness in a ribbed pork carcass is primarily evaluated at the _____ rib.

15. Ideal color of fresh pork is _____.

16. Presence of a break joint in a lamb carcass indicates the lamb was _____ of age or less.

17. The most expensive cuts of lamb are located _____.

18. True or False. Meat judging reasons are always given orally.

19. True or False. All pork slaughter plants utilize the USDA standard yield and quality grading system.

20. True or False. Meat judging reasons should be delivered or written with comparisons of the top, middle, and bottom pairs.

Objectives

After completing this chapter, students should be able to:

▶ Describe an ideal market hog, steer, and lamb

▶ Place classes of market hogs, steers, and lambs

▶ Describe ideal male and female breeding stock

▶ Place classes of male and female breeding stock

Key Terms

blind nipples	calf-kneed	pigeon-toed
bowlegged (front legs)	cowhocked	pin nipples
bowlegged (rear legs)	finish	posty
buck-kneed	flat nipples	sickle-hocked
bucks	knock-kneed	splay-footed

Career Focus

Jenessa's livestock marketing business began during her stint as a livestock judging-team member in high school. Her coach encouraged her to become familiar with the terminology associated with the purebred livestock industry through practicing reasons. At the time Jenessa thought she was learning a new language—which in fact she was. In college, Jenessa majored in agricultural business with an emphasis in marketing and interned at two Web-based marketing firms. She even made time to compete at her school's collegiate livestock judging team. During one of her judging trips Jenessa met the owner of a small, start-up company that built Web sites for livestock producers. This meeting resulted in Jenessa's current position where she spends her days photographing livestock, designing Web sites, and developing sales fliers and catalogs for purebred livestock producers to aid in their marketing efforts.

INTRODUCTION

Livestock judging is the art and science of evaluating the phenotype of live animals. Phenotype gives indications of the utility of livestock on a farm, ranch, or feedlot that performance data cannot. For example, performance data give no clues to the structural correctness of an animal, which may affect the longevity in a breeding herd. Ideally, a combination of performance data and visual appraisal should be used to make placing decisions. Unfortunately, data are not always available. Often, visual evaluation is the only method available for livestock selection.

MARKET ANIMAL EVALUATION

Market animals are often evaluated, bought, and sold solely by phenotype. Livestock buyers make decisions based on the apparent amount of salable muscle in a live animal. Many individuals have perfected their abilities to estimate the relative amounts of muscle and fat. Livestock buyers make a living based on this ability. Judging any market animal relies on three parameters: muscle, fat, and structural correctness. The following is a discussion of places to look on live market animals for indications of these parameters.

BEEF

As with meats judging, proper terminology is essential to describing a market steer or heifer. Memorize the parts presented in figure 16-2.

Evaluating muscle content in market steers and heifers is not difficult. As with all market animals, the highest-priced cuts of meat originate from the rear half of the animal. See figure 16-3. Therefore, select steers or heifers that are widest over the top, rump, and rear quarter.

FIGURE 16-1

Livestock are typically evaluated in classes of four.

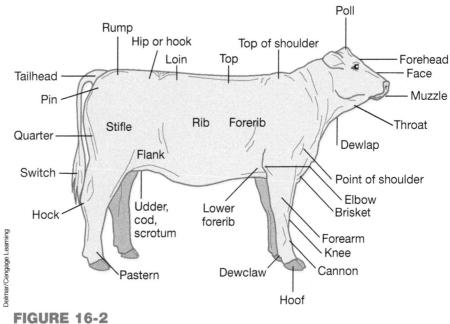

FIGURE 16-2

Parts of a beef animal.

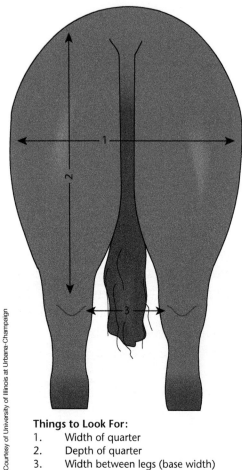

Courtesy of University of Illinois at Urbana-Champaign

Things to Look For:
1. Width of quarter
2. Depth of quarter
3. Width between legs (base width)

FIGURE 16-3

Dimensions of muscling on the rear view of beef cattle.

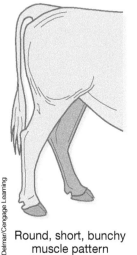

Delmar/Cengage Learning

Round, short, bunchy muscle pattern

Longer, smoother muscle pattern

FIGURE 16-4

Two types of muscle design in beef cattle.

Muscle pattern and design are also important. Muscle should be long, smooth, and thick, not short, bunchy, and round. See figure 16-4.

Terms used to describe muscle content in market cattle follow.

DESIRABLE	UNDESIRABLE
Thicker made	Narrower made
Wider based	Narrower based
Heavier muscled	Lighter muscled
Longer, smoother muscle design	Shorter, tighter muscle design
Thicker quartered	
Deeper quartered	Shallower quartered
Wider topped	Narrower topped
Thicker stifled	
Pushed more stifle on the move	
Wider rumped	Narrower rumped
More total muscle mass	
More muscle dimension	
Rail a carcass with a larger ribeye	

Fat thickness in market steers and heifers must be adequate, but not excessive. Remember, a certain amount of exterior fat must be present to assure adequate marbling. Too much exterior fat results in poor yield grading, low-cutability carcasses. In a live steer or heifer, fat may be most easily evaluated by observation of a deep, full brisket, an overall smooth appearance, and patches of fat beside the tailhead. Some judging contests allow market cattle to be handled to determine correct **finish** (fat cover). Experienced judges handle cattle by placing four fingers together and running them along the upper and lower rib cage. See figure 16-5. Fat cover will squish out from under the fingers, while unfinished or underfinished cattle will feel hard and bony to the touch. Some cattle will have adequate fat cover over the upper rib, but will be free of finish over the lower rib. These cattle are referred to as unevenly finished.

The following are terms used to describe finish on market cattle.

DESIRABLE	UNDESIRABLE
More ideally finished	Wastier
Smoother in his/her finish	Patchy, uneven finish
Should rail a carcass with	
—a higher lean to fat ratio	
—a more desirable yield grade	
More apt to grade choice	
Trimmer, cleaner patterned	
Trimmer middled	Wastier middled

Structural correctness incorporates frame size, outline, and balance, as well as feet and legs. The ideal market steer should be moderately framed, well balanced, and boxy in appearance with a strong,

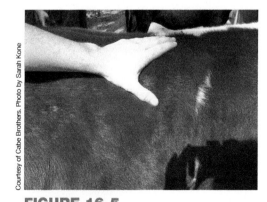

Courtesy of Cabe Brothers. Photo by Sarah Kone

FIGURE 16-5

Determine finish on steers by handling over the ribeye.

straight topline. Feet and legs should be spaced evenly under each corner of the box. The rump should be level. The neck should be long and clean, and the point of the shoulder should be neatly laid-in. The ideal market steer should be long bodied, deep bodied, and have good spring of rib. Use the steer in figures 16-6A and 16-6B as your example of an ideal market steer and compare all other market cattle to that mental image.

The description of feet and legs requires knowledge of correct skeletal design and the ability to recognize when that design is not quite right. From the side, rear legs can be **posty** (too straight), or **sickle-hocked** (too curved). From the rear, legs can be bowed in (**cowhocked**) or out (**bowlegged**). Front legs can also have similar problems. When viewed from the side, front legs that are too straight are called **buck-kneed**. Front legs with too much curvature are called **calf-kneed**. When viewed from the front, legs can be **bowlegged** (bow out), **knock-kneed** (knees too close together), **splay-footed** (feet turned out), or **pigeon-toed** (feet turned in). See figure 16-7 for pictures of common rear leg faults and figure 16-8 for common front leg faults.

Photo courtesy of Judging101.com

FIGURE 16-6A

A very stylish, thick market steer on a side view.

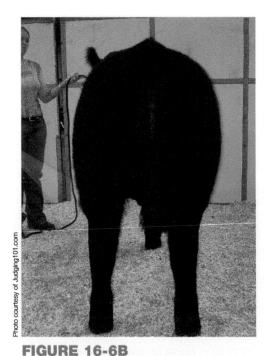

Photo courtesy of Judging101.com

FIGURE 16-6B

The same steer displaying his natural width from behind.

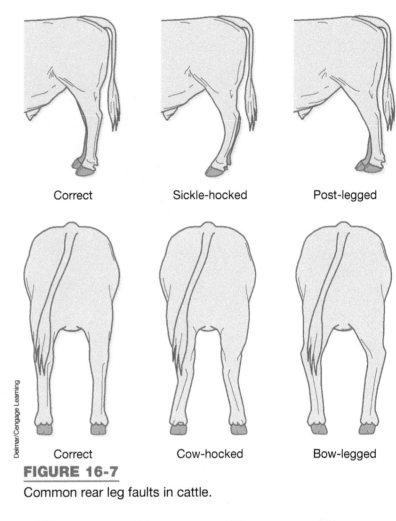

Correct   Sickle-hocked   Post-legged

Correct   Cow-hocked   Bow-legged

FIGURE 16-7

Common rear leg faults in cattle.

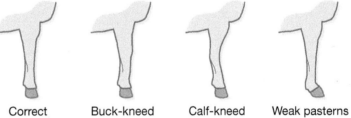

Correct   Buck-kneed   Calf-kneed   Weak pasterns

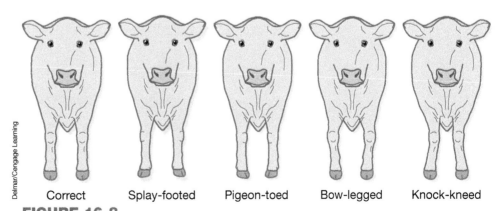

Correct   Splay-footed   Pigeon-toed   Bow-legged   Knock-kneed

FIGURE 16-8

Common front leg faults in cattle.

The following are terms used to describe structural correctness in market cattle.

DESIRABLE	UNDESIRABLE
FRAME/OUTLINE	
Better balanced	Off balance
More eye appealing	
Fresher appearing	Staler appearing
Moderate framed	Smaller or larger framed
Straighter lined	Slack framed
More ideal in his maturity pattern	Earlier or later maturing
Stronger topped	Weaker topped
More structurally correct	Structurally incorrect
Fault-free	Ill-structured
Longer bodied	Shorter bodied
CAPACITY	
Bigger volumed	More constricted
More arch and spring of rib	Flatter ribbed
Deeper ribbed	Shallower ribbed
Deeper flanked	Shallower flanked
FRONT ENDS	
More extension through his/her front end	Shorter fronted
Smoother, tighter shouldered	Open shouldered
Laid-in more neatly about the shoulder	Coarser fronted
Longer, leaner neck	
Tighter fronted	
RUMP	
Squarer, leveler rump	Narrow, steep rump
Wider/longer from hooks to pins	Narrower from hooks to pins
Wider rump	Narrower rump

DESIRABLE	UNDESIRABLE
FEET/LEGS/MOVEMENT	
Heavier boned	Finer boned
More desirable set to the hock	Posty, sickle-hocked
Stronger pasterns	Weaker pasterns
Stood squarer when viewed from the rear/front	Cowhocked, toes out
Stood/tracked wider behind	Narrower tracking/based
Longer strided	Shorter strided
Moved freer and easier	Shorter strided
Moved with more strength of top	Roached his/her top on the move
Moved with more level-ness of rump	Dropped his pins on the move
GENERAL	
More progressive	Conventional
Performance oriented	Lacks performance
Growthier	
More dimensional	
Producer-Oriented	

Sample reasons for market steers
Placing: 1–2–3–4

In my top pair of heavier-muscled steers, I preferred 1 to 2 as 1 was a trimmer, cleaner-patterned steer that was more ideal in his finish. 1 was a longer-necked, tighter-shouldered steer that was stronger in his top and more nearly level through his rump. He also handled more smoothly over his lower rib and should be more apt to rail a carcass reaching the choice grade. I realize that 2 was a heavier-boned, deeper-bodied, more ruggedly designed steer.

Nonetheless, it was 2 over 3 in my middle pair. 2 simply overpowered 3 in total muscle dimension.

2 was a thicker-topped steer that carried his muscle mass into a wider, more expressively muscled rump and a deeper, thicker quarter. In addition, 2 was a wider-tracking steer that pushed more stifle on the move. He should rail a carcass with a larger ribeye. I must admit that 3 was a trimmer-middled, tighter-fronted steer, but he lacked the sheer muscle volume of my top pair.

Moving to my bottom pair of lighter-muscled, narrower-tracking steers, I preferred 3 to 4 as 3 was a tighter-framed, more structurally correct steer. 3 was a longer-necked, tighter-shouldered, stronger-topped steer that had a more desirable set to his hock. Thus, 3 moved with a longer, more ground-covering stride.

I fully realize that 4 was more performance oriented, deeper ribbed, and deeper flanked. But, 4 was placed last in the class as he was the most structurally incorrect, off-balance steer. He lacked the overall style and symmetry of the steers placed above him. Thank you.

SWINE

Figure 16-9 illustrates the parts of a market hog. Study them with care.

Muscle evaluation of market hogs relies on an estimate of thickness. However, thickness caused by fat looks nearly the same as thickness caused by

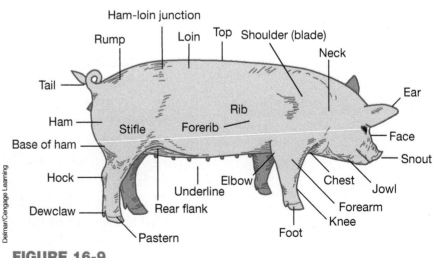

FIGURE 16-9

Parts of a hog.

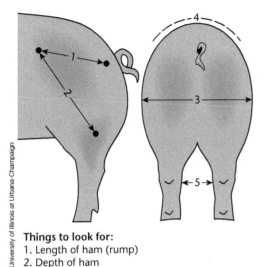

Things to look for:
1. Length of ham (rump)
2. Depth of ham
3. Width through center of ham
4. Correct turn over the top
5. Width between hind legs (base width)

FIGURE 16-10

Dimensions of muscling in a hog's ham.

Courtesy of University of Illinois at Urbana-Champaign

muscle. It takes a practiced eye to tell the difference. Often, the leanest, heaviest-muscled pig will not be the widest-topped pig in the class. Muscle can often be evaluated more accurately by observing width between the front legs and rear legs while the pig is walking. Pigs with a wider stance usually have more true muscle volume than pigs whose legs are set close together.

Places to look for muscle on a market hog naturally include the loin, rump, and ham. See figure 16-10. Hogs thick in these areas containing high-priced cuts will probably have wide shoulders and chests also.

The following are terms used to describe muscle content in market hogs.

DESIRABLE	UNDESIRABLE
Wider based/tracking	Narrower based/tracking
Heavier muscled	Lighter muscled
More expressively muscled	Flatter muscled
More natural turn to his/her top	
Pushes more stifle on the move	Narrower stifled
More total muscle mass	Less total muscle mass
Wider rumped	Narrower rumped
More muscle dimension from stifle to stifle	
More flair and dimension through the lower 1/3 of the ham	
Should rail a carcass with a higher lean to fat ratio	
Should rail a carcass with a larger loineye	
Should rail a carcass with a higher percent lean	

The ability to visually evaluate fat on market hogs is probably the most important and difficult skill learned by hog judges. Market hogs are not handled to evaluate fat cover, so visual clues are the only indication of relative fatness.

Fortunately, visual clues to fatness exist. A square shape over their top when viewed from the rear identifies pigs carrying too much fat. The shoulder blade will be made almost invisible by a thick fat layer. Additionally, pigs carrying too much fat will appear sloppy in the jowl, at the base of the ham, and in the flank. A roll of fat may appear in the elbow pocket when the pig's front leg is extended to the rear.

An exceptionally lean pig will show a definite ham-loin junction and a more rounded shape to the top when viewed from the rear. The shoulder blade will be obvious when the pig moves. A lean pig will appear trim in the jowl, flank, and at the base of the ham. See figure 16-11 for a very lean pig.

In order to provide pork with enough fat to provide a pleasant eating experience, market pigs should not be too lean. In addition, ultra-lean pigs often grow slowly. When judging market hogs, it is best to avoid extremes in fat cover.

The following are terms used to describe relative leanness in market hogs.

Photo by Frank Flanders

FIGURE 16-11

This gilt is not structurally ideal, but she is extremely lean.

DESIRABLE	UNDESIRABLE
LEANER DESIGNED	
Raw made	
Leaner topped	Plainer topped
Cleaner over the loin edge	Squarer over the loin edge
More natural shape his/her top	
Trimmer in his/her lower one third	Wastier in his/her lower one third
Showed more blade action on the move	
Should rail a carcass with a greater lean value	

Structural correctness of market hogs also deserves a great deal of attention. A correctly designed pig should be large framed, long and deep bodied, level topped, and naturally wide from front to rear. See figures 16-12A and 16-12B to help visualize an ideal market hog.

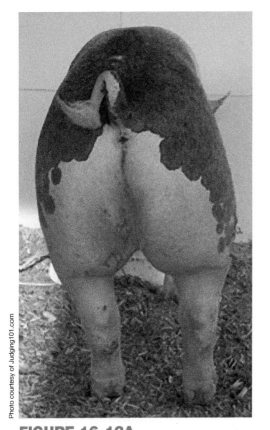

Photo courtesy of Judging101.com

FIGURE 16-12A

Note the natural base width and muscle dimension of this barrow from the rear.

Photo courtesy of Judging101.com

FIGURE 16-12B

A stout, powerful, easy feeding market barrow from the side.

Feet and legs should show a great deal of flex at all joints when the pig moves. See figure 16-13 for pictures of common leg structure faults.

The following are terms used to describe structural correctness in market hogs.

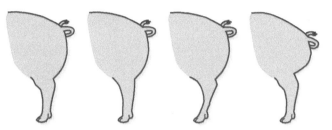

| Correct | Posty, straight in the back | Sickle-hocked | Steep-rumped |

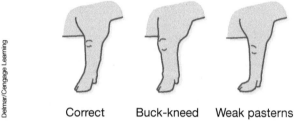

| Correct | Buck-kneed | Weak pasterns |

Delmar/Cengage Learning

FIGURE 16-13

Common leg faults of swine.

DESIRABLE	UNDESIRABLE
FRAME/OUTLINE	
Larger framed	Smaller framed
Bigger outlined	Smaller outlined
Longer sided	Shorter sided
Leveler topped	Higher topped
Better balanced	More off balance
MORE FUNCTIONAL	
Looser made	
More rugged	More frail
Sounder made	Poorer structured
More durable	Less durable
CAPACITY	
Wider sprung	Narrower made
Deeper ribbed	Shallower ribbed

Animal Science Facts

Individual animals in a judging contest may have identification points that should be referenced in a set of reasons. For example, a heifer that has ripped her ear tag out leaving a split in the ear could be referred to as "the split-eared heifer." Likewise, males and females should be differentiated in a class of mixed market animals. Color patterns that differ from the norm should also be identified. Pigs with identical notches in the right ear can be identified as "potential litter-mates" in a swine class. Using identification points helps the listener better follow a set of reasons and helps make the contestant seem more observant and knowledgeable about the class.

DESIRABLE	UNDESIRABLE
Deeper chested	Tighter ribbed
Wider through his/her chest floor	Narrower chested
Wider based	Narrower based
Bolder sprung	
Deeper flanked	Tighter flanked
More expanded through his/her rib	Constricted in his/her rib
Looser middled	Tighter middled
RUMP	
Longer rumped	Shorter rumped
Leveler rumped	Steeper rumped
Leveler in his/her rump design	
Wider rumped	Narrower rumped
Squarer rumped	
FEET/LEGS/ MOVEMENT	
More confinement adaptable	Less confinement adaptable
More curvature to his/ her knee	Buck-kneed
More desirable slope to his/her shoulder	Straighter shouldered
More flex of hock	Straighter hocked
More cushion in his/her pastern	
More animated in his/ her movement	
Looser strided	Tighter wound
Longer strided	Shorter strided
Wider tracking	Narrower tracking
More flexible	Tighter made

DESIRABLE	UNDESIRABLE
Heavier boned	Finer boned
Stood on greater substance of bone	
GENERAL	
Growthier	
Higher performing	Poorer performing
More functional	
More producer oriented	
More complete	
More fault-free	

Sample market hog reasons

Placing: 2–3–4–1

In my top pair, I preferred 2, the blue-butt barrow over 3, the red gilt, as 2 was a more producer-oriented, confinement-adaptable barrow that should go to the rail and hang a carcass with a higher-percent muscle. 2 was a wider-sprung, deeper-ribbed barrow that was thicker over his top, through his rump, and showed more natural thickness from stifle to stifle. In addition, he stood on more substance of bone and moved with more flex and animation to his hock. I grant that 3 was trimmer through her lower one third.

Nevertheless, it was 3 over 4 in my middle pair. 3 was leaner-made, more expressively muscled gilt that should rail a higher-cutability carcass. 3 showed more blade action on the move, displayed a more natural turn to her top, and was more expressive through the center and lower portions of her ham. I realize 4 was more nearly level through his rump and moved with more curvature and cushion to his knee and pastern, but he was plainer in his top and narrower based.

4 placed over 1 in my bottom pair of black barrows. 4 more closely followed my top barrow in his kind being a looser-designed, bigger-scaled, wider-made barrow from end to end. 4 was a wider-chested, deeper-ribbed barrow that moved out with a longer, looser stride. Additionally, he was a wider-topped, wider-rumped barrow that stood on a greater diameter of bone.

Granted, 1 was leaner over his loin edge and would hang a carcass with less fat trim. However, 1 was the poorest-performing, narrowest-made, lightest-muscled, shallowest-bodied pig in the class that had the least to offer the performance-minded producer. Thank you.

LAMBS

The parts of a market lamb are shown in figure 16-14.

Muscle thickness in lambs is easiest to detect through handling. Many judging contests allow contestants to handle all lambs. A heavily muscled lamb will visually appear wide from directly behind the shoulders, through the rump, and stand naturally wide when viewed from the rear. See figure 16-15. When

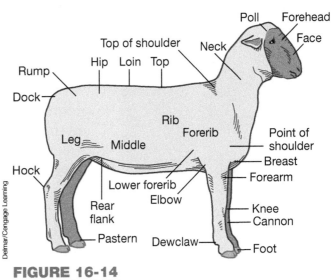

FIGURE 16-14

Parts of a sheep.

Look for width—
1. At the center of leg
2. At the dock
3. Between hind legs
4. Over back and loin

Look for—
5. Long rump
6. Long, bulging stifle
7. Depth of leg
8. Width at chest
9. Heavy bone

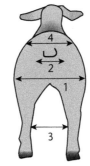

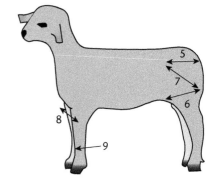

FIGURE 16-15

Dimensions of muscling in a market lamb.

FIGURE 16-16

With fingers extended and close together, try to estimate the depth of fat squishing out from under your fingers as you handle the lamb's topline.

FIGURE 16-17

Compare length of loin among lambs in the class. The most valuable cuts are found from the last rib back.

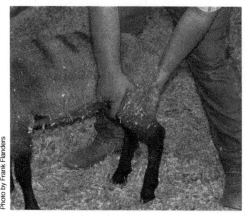

FIGURE 16-18

Leg muscle dimension can be determined by grasping the leg as close to the lamb's flank as possible.

handled, a heavily muscled lamb should feel wide and thick over the top directly behind the shoulders. The width should be maintained through the loin and get progressively wider through the rump. The rump and leg should feel thick, full, and square. When grasped directly behind the last rib, the loin should be deep, as well as wide. The length of loin can be evaluated by using the hands to measure the distance between the last rib and the hip.

Each lamb should be handled with a predetermined routine so differences among lambs can be remembered. The four fingers should be extended and kept together. See figure 16-16. First, handle the topline (with one hand) from behind the shoulder, progressing to the rump. Next, with both hands, determine the depth and length of loin. See figure 16-17. Last, with both hands, grasp the leg as close to the lamb's flank as possible. Touch the tips of your fingers on the inside of the leg and note the distance between the thumbs on the outside of the leg. This distance can be used to compare leg muscle among lambs. See figure 16-18.

The following are terms used to describe muscle content in market lambs.

DESIRABLE	UNDESIRABLE
Heavier muscled	Lighter muscled
Thicker made	
More expressively muscled leg	Narrower, lighter muscled leg
More dimensional leg	
Heavier muscled hindsaddle	Lighter muscled hindsaddle
Handled with a longer loin/deeper loin/ thicker rump/heavier muscled leg	
Should hang a carcass with a higher percent hindsaddle	
Should rail a carcass with a heavier muscled leg	
Met my hand with more width and muscularity directly behind the shoulder	

Fat determination can also be accomplished by handling a market lamb. Sometime during the handling process, gauge the amount of fat that squishes out from under the fingers when a lamb is handled over the upper and lower rib. See figure 16-19. Carefully pinching the rear flank of the lamb for fullness can also give a good approximation of the relative fat cover over the rib. A lamb that has a full, fat-filled flank is probably fat over the rib. Remember, ideal fat cover is 0.10 to 0.15 inches. Anything over 0.25 inches is excessive. Accurately determining the amount of fat on market lambs takes considerable practice.

Fat can also be estimated by visual appraisal. Fat lambs will appear smooth, while especially trim lambs will appear lumpier.

The following are terms used to describe fat differences in market lambs.

Photo by Frank Flanders

FIGURE 16-19

Sometimes lambs handle with more finish over their rib than over their topline.

DESIRABLE	UNDESIRABLE
Trimmer made	Wastier
Trimmer, cleaner patterned	Overfinished
Cleaner conditioned	Heavier conditioned
Harder handling	Softer handling
Handled with more trimness over the upper and lower rib/at the 12th rib	
Should hang a trimmer carcass	Should hang a wastier carcass
Should hang a higher-cutability carcass	

Structure of market lambs is most similar to that of cattle. Therefore, the terminology used is also similar. An ideal market lamb should be wedge shaped. The front end should be trim, the neck long and lean, and the shoulders tight and neatly laid-in. The topline and dock should be straight and level when viewed from the side. The body should be deep, but not wasty. The loin and the leg should be much wider than the front end. See figures 16-20A and 16-20B for an example of a very good market lamb.

Photo courtesy of Jenny Dietrich

FIGURE 16-20A

A nicely balanced, meaty market lamb from the side.

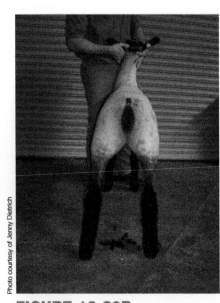

Photo courtesy of Jenny Dietrich

FIGURE 16-20B

This lamb gives every indication of muscularity when viewed from the rear.

Feet and legs should be squarely placed with the front legs spaced slightly closer together than the rear legs. Bone should be heavy, but not coarse. Feet and leg faults are similar to those described for cattle.

The following are terms used to describe structural correctness in market lambs.

DESIRABLE	UNDESIRABLE
FRAME/OUTLINE	
Better balanced	Off balance
More eye appealing	
Larger framed	Smaller framed
More size and scale	
Tighter framed	Slack framed
Later-maturing	Earlier-maturing
Stronger topped	Weaker topped
More structurally correct	Structurally incorrect
Fault-free	Ill-structured
Longer bodied	Shorter bodied
CAPACITY	
Wider sprung	Narrower made
Higher capacity	Tighter ribbed
Trimmer middled	Wastier middled
FRONT ENDS	
Fresher appearing	Staler
More extension through his/her front end	Shorter fronted
Smoother, tighter shouldered	Open shouldered
Laid-in more neatly about the shoulder	Coarser fronted
Longer, leaner neck	Pelty about the neck
Tighter fronted	
Smoother neck-shoulder junction	Ewe necked

DESIRABLE	UNDESIRABLE
DOCK	
Squarer, leveler dock	Narrow, steeper dock
Wider/longer dock	
Wider rump	Narrower rump
FEET/LEGS/ MOVEMENT	
Heavier boned	Finer boned
More desirable set to the hock	Posty or sickle-hocked
Stronger pasterns	Weaker pasterns
Stood squarer when viewed from the rear	Cowhocked
Stood wider behind	Stood narrower behind
Stood squarer in his/her foot placement	Cowhocked, bowlegged
GENERAL	
More progressive	Conventional
Performance oriented	Lacks growth and performance
Growthier	
More dimensional	
Stouter-made	
More upstanding	

Sample market lamb reasons

Placing: 4–3–2–1

In my top pair of wethers, I place 4 over 3 as 4 was a trimmer-handling, cleaner-patterned lamb that should go to the rail and hang a higher-cutability carcass. 4 was longer in his neck, more neatly laid-in at the point of his shoulder, and handled harder over his top. I admit 3 was a heavier-muscled lamb that was deeper and wider in his loin, but he was a softer-handling, wastier-fronted lamb.

Nonetheless, I preferred 3 to 2 in my middle pair because 3 outdistanced 2 in terms of balance and muscle. 3 was a more upstanding, eye-appealing lamb that was longer in his hindsaddle. In addition, he handled with a deeper loin, a wider, squarer dock, and a more expressively muscled leg. He should rail a heavier-muscled carcass. I grant 2 was more correct in his finish, but he was an off-balance, coarser-fronted lamb that lacked the muscle dimension of my top pair.

Moving to my bottom pair of ewe lambs, I placed 2 over 1 because 2 was more correctly finished and should rail a trimmer, higher-cutability carcass. 2 handled harder over her top and was trimmer in her flank.

I realize 1 was a taller-fronted, more upstanding lamb, but she was the wastiest, heaviest-conditioned, lightest-muscled lamb that should hang the least profitable carcass in the class. Thank you.

BREEDING ANIMAL EVALUATION

Evaluation of breeding livestock often includes performance data. However, when data are not available, breeding livestock should be selected based on their apparent ability to be the parents of ideal market animals. Therefore, muscle, fat, and structural correctness should receive significant attention, with structure moving toward the top of the priority list. In addition, breeding soundness, body capacity, and visual clues to performance must be included in the evaluation.

BEEF

Breeding cattle should have some of the same characteristics as an ideal market steer. The basic structure should be about the same, with small refinements discussed in a set of reasons. For example, instead of focusing solely on the potential carcass cutout of the animal in question, the animal's structure, body capacity, femininity, masculinity, and performance should be emphasized. Breeding cattle must be relatively fault-free in their makeup and designed with added body depth and volume to perform in all management conditions—even when feed resources are limited. Most judging classes of breeding cattle will be heifer classes, so most of the following discussion will be focused on females. However, classes of bulls are sometimes evaluated in judging contests.

Heifers

See figure 16-21A and 16-21B for an exceptionally good heifer.

As with market cattle, muscle dimension and pattern are important to breeding heifers. Use the same terms used to describe muscle volume in market cattle. However, carcass terms should be avoided.

Relative body condition is used to describe heifers, but for a different reason than with market cattle. Condition can give an indication of the ability of a heifer to survive and thrive in various management situations. Heifers should be in good condition, but not exceptionally thin or fat.

The following are terms used to describe body condition in beef heifers.

FIGURE 16-21A

A deep bodied, broody Angus heifer.

FIGURE 16-21B

This heifer is very correct and adequately thick when viewed from behind.

DESIRABLE	UNDESIRABLE
Easier keeping	Harder doing
Easier fleshing	Harder fleshing
	Heavier conditioned

Structure is more critical to breeding heifers than market cattle. Correctly structured individuals (especially on their feet and legs) are less apt to be culled for unsoundness and, thus, should have a longer life in the breeding herd. Structural terms are the same as those used for market cattle.

Femininity is important to heifers. Femininity is generally recognized by a long, clean, smooth front end.

The following are terms used to describe femininity in beef heifers.

DESIRABLE	UNDESIRABLE
Broodier	
More feminine fronted	Coarser fronted
More angular	
More stylish	
Longer, more feminine headed	

Visual indications of performance include overall body length, depth, and capacity, as well as relative size and weight in comparison to other heifers in the class.

The following are terms used to describe visual indications of performance in beef heifers.

DESIRABLE	UNDESIRABLE
More three-dimensional	
Bigger volumed	Had less volume
Broodier	
Roomier	

DESIRABLE	UNDESIRABLE
Higher weight per day of age	Lower weight per day of age
More producer oriented	
More performance oriented	
More arch and spring of rib	Constricted in her rib
More capacious	Less capacious
More dimension through the center of her rib	
Longer ribbed	Shorter ribbed
Growthier	Lacked doability

The following are general introductory terms used to describe beef heifers.

DESIRABLE	UNDESIRABLE
Better combination of performance and eye appeal	Poorest combination of performance and eye appeal
Better combination of structure and performance	Poorest combination of structure and performance
More progressive	Least progressive

Sample Angus heifer reasons
Placing: 3–2–4–1
In my top pair, I placed 3 over 2 because 3 better combined femininity and performance into a broodier, more producer-oriented package. 3 was a cleaner-fronted, longer-necked heifer that was deeper ribbed, heavier muscled, and stood on more substance of bone. In addition, 3 appeared to have a higher weight per day of age. I admit that 2 was more structurally correct, more nearly level from her

Animal Science Facts

The discussion of testicular development may seem in poor taste when giving reasons on a class of bulls, rams, or boars. However, testicle size and structure are economically important traits in terms of a sire's fertility. Larger testicles produce more sperm. Therefore males with larger testicles should have a better chance of impregnating females. Collection of large volumes of sperm is also desirable for artificial insemination. Lastly, large testicle size is an indication of earlier sexual maturity for a sire's daughters.

hooks to her pins, and moved out with a longer, more ground-covering stride.

In my middle pair, I preferred 2 to 4 as 2 was a better-balanced, more upstanding, later-maturing heifer. 2 was laid-in more neatly at the point of her shoulder, was stronger topped, leveler rumped and stood with a more correct set to her hock. Also, 2 tracked wider and squarer both coming and going. I grant 4 was deeper through her forerib and wider at her pins, but she simply lacked the balance, femininity, and structural correctness of 2.

Moving to my bottom pair, it was 4 over 1. 4 was an easier-keeping, bigger-volumed, more performance-oriented heifer that was more rugged in her design. She was a more 3-dimensional heifer that showed more arch and spring of rib, more depth of flank, was more ideal in her condition, and heavier boned.

I realize 1 was more refined and feminine fronted, but she lacked the frame, muscle, and doability to place any higher in the class. Thank you.

Bulls

Most terms used to describe bulls can be gleaned from the terminology lists already presented. However, bulls should be described using terms that indicate the bull's muscle, masculinity, structural correctness, and the kind of progeny he will most likely sire.

The following are terms used specifically to describe bulls.

DESIRABLE	UNDESIRABLE
Stouter	
More powerful	
Greater scrotal circumference	Less scrotal circumference
More masculine	
More athletic	
More agile	Restricted in his movement

DESIRABLE	UNDESIRABLE
Should sire calves with: added growth and performance added muscle volume added frame added length of body	

SWINE

Judging breeding swine is very similar to judging market swine. The only difference is that structure, frame size, and body capacity should receive more emphasis than in market classes. Muscle and leanness are also very important.

Gilts

See figure 16-22A and 16-22B for the ideal breeding gilt. Good replacement gilts should be large framed, long bodied, and late maturing with adequate muscle and bone. High-volume, wide-based, deep-ribbed gilts that move easily are preferred. Replacement

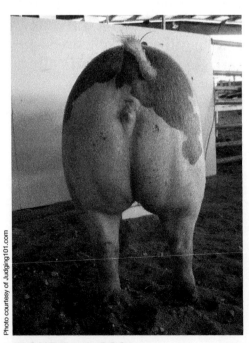

FIGURE 16-22A

This gilt is wide based and very muscular without being coarse and terminal in her appearance.

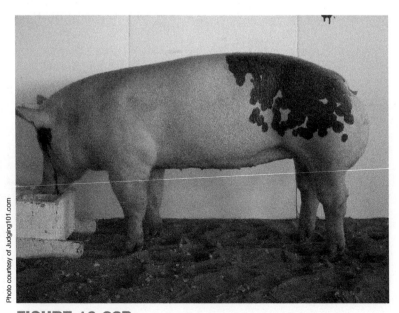

FIGURE 16-22B

A deep ribbed, ruggedly designed crossbred breeding gilt.

gilts must also have an underline with at least six functional nipples per side. See figure 16-23. **Blind nipples** (nonfunctional), **flat nipples** (functional, but not prominent), and **pin nipples** (small, undeveloped nipples) are discriminated against. See figure 16-24. In most cases, judging contest contestants are told not only to judge underlines as sound (having at least six functional nipples), but also to discuss differences in their reasons. Underline quality should also be examined in boar classes (particularly boar classes of maternal breeds).

Terms used to describe muscle content and fat cover are the same as those used for market swine. Because these are breeding animals, carcass terms should not be used.

Photo by Frank Flanders

FIGURE 16-23

Breeding gilts should have at least six functional nipples per side.

Delmar/Cengage Learning

Ideal

Nipples 2 and 3
are flat or blind

Pin nipple between
nipples 3 and 4

FIGURE 16-24

Common underline faults in swine.

The following are some additional terms used to describe body volume and performance in breeding gilts.

DESIRABLE	UNDESIRABLE
More three-dimensional	
Bigger volumed	Had less volume
Roomier middled	
Higher weight per day of age	Lower weight per day of age
More performance oriented	
More internal capacity	Less capacious
More natural width through the center of her rib	
Deeper sided	

Remaining terms concerning structure, as well as feet and legs, can be gleaned from the market terminology lists.

The following are terms used to describe femininity and underlines in gilts.

DESIRABLE	UNDESIRABLE
Longer fronted	shorter fronted
More feminine fronted	
Higher-quality underline	Poorer-quality underline
More desirable teat spacing	Unevenly spaced teats
More refined underline	Coarser in her teat quality
More numerous underline	Less numerous underline
More prominent underline	Possessed a pin nipple
	Was blunt in her teats

Sample Hampshire gilt reasons
Placing: 4–2–1–3
I started this class with 4 over 2 as 4 was a bolder-sprung, roomier-middled, heavier-muscled, more structurally correct gilt. 4 was wider based, deeper ribbed, deeper flanked, and was heavier muscled from blade to ham. In addition, 4 was looser in her hip and moved with a longer, more ground-covering stride. I grant that 2 was taller fronted, longer necked, and especially leaner over her top and through her lower one third, but she was simply narrower made from end to end.

Nonetheless, I placed 2 over 1 in my middle pair as she was a larger-framed, later-maturing gilt that appeared to have a higher weight per day of age. 1 was leaner at her blades, taller fronted, longer necked, and especially leaner in her lower one third. I realize that 1 was the wider-based gilt, but she was a much earlier-maturing female that lacked the carcass composition of my top pair.

In my bottom pair, I preferred 1 to 3 as she more closely followed my first place gilt. She was a more producer-oriented gilt with more volume, rib, and capacity than 3, displaying a bolder rib shape and more natural width at the ground. 1 tracked wider both coming and going while being more numerous and prominent in her underline.

I grant 3 was a functional gilt in terms of her mobility, but she was simply the narrowest-based, shallowest-ribbed, poorest-performing gilt in the class. Thank you.

Boars

Except for the following few terms, most terminology is the same as that used for market hogs and gilts.

DESIRABLE	UNDESIRABLE
More powerful	
Heavier structured	
More testicular volume	Less testicular volume

DESIRABLE	UNDESIRABLE
Potential to sire pigs with: added growth and performance; additional leanness and/ or muscle; more soundness; more confinement adaptability	

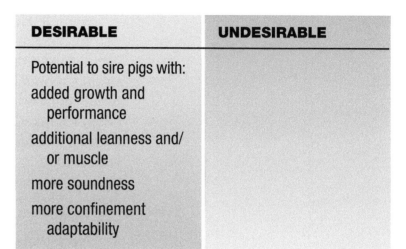

SHEEP

The evaluation of breeding sheep is similar to that of market lambs. However, structure, volume, performance, and feet and legs should all receive a somewhat higher priority.

Ewes

Ewe classes are judged as lambs (less than one year) or yearlings (greater than one year). Maturity pattern should be emphasized more in ewe lambs than in yearling ewes. The ideal ewe should have the same general characteristics as the ideal beef heifer. She should be trim and feminine fronted but thick, deep, large volumed, and heavily muscled. Feet and legs should be correctly placed and sound. See figure 16-25 for an ideal ewe.

Photo courtesy of Linde's Livestock Photos

FIGURE 16-25

An ideal breeding ewe.

Muscle and fat evaluation for breeding sheep are normally made without the benefit of handling. Judging contestants must evaluate carcass traits based on the appearance of thickness for muscle content and smoothness for fat evaluation.

The following are terms used to describe muscle volume in breeding sheep that have not been handled.

DESIRABLE	UNDESIRABLE
Longer, smoother muscle design	
Thicker made	
More expressively muscled leg	Narrower, lighter-muscled leg

Body condition is important to breeding ewes for the same reasons as it is for breeding heifers.

The following are terms used to describe fat differences in breeding sheep that have not been handled.

DESIRABLE	UNDESIRABLE
Trimmer made	Wastier
Trimmer, cleaner patterned	Overfinished
Cleaner conditioned	Heavier conditioned
More ideal in her condition	
Trimmer breasted	Wastier in her breast
Easier keeping	Harder doing
Easier fleshing	

Most structure terms are the same as those used for market lambs.

The following are additional structure terms to be used in breeding sheep reasons.

DESIRABLE	UNDESIRABLE
MOVEMENT	
Moved freer and easier	Stiffer strided
Moved with more strength of top	
Moved with more levelness of dock	
Longer strided	Shorter strided
Truer tracking	Narrower tracking
Sounder footed	

In addition to the capacity terms used for market lambs, the following terms can be used to describe volume and capacity in breeding sheep.

DESIRABLE	UNDESIRABLE
More capacious	Shallower made
More internal volume and dimension	
More arch and spring of rib	Fatter ribbed
Bolder spring of rib	Constricted in her forerib
More three-dimensional	

Ideally, fleeces on breeding sheep should be dense and tight with few black fibers. Fleece quality should be emphasized when judging breeds known for their wool production. Fleece quality differences should be discussed using the following terms.

DESIRABLE	UNDESIRABLE
Tighter denser fleece	More open fleece
Fleece with a finer crimp	Cottony fleece
Freer from black fiber	Fleece with more black fiber

Sample Dorset ewe reasons

Placing: 2–1–4–3

In my top pair, I easily placed 2 over 1 as she was a more progressive ewe in terms of frame, muscle, and volume. She was a trimmer, cleaner-patterned, taller-fronted ewe that was thicker topped, deeper ribbed, and tracked wider behind. I concede 1 was heavier boned.

Moving to my middle pair of similarly designed, more conventional ewes, I preferred 1 to 4 as she was a cleaner-fronted, sounder-footed ewe. 1 was longer and leaner necked, laid-in neater at her shoulder, and was smoother at the neck-shoulder junction. Additionally, she was stronger and straighter on her pasterns and tracked with more ease and agility. I grant 4 was more nearly level through her dock and possessed a tighter, denser fleece.

Nevertheless, I confidently placed 4 over 3 in my bottom pair as she was a bolder-sprung, more capacious, more performance-oriented ewe. 4 showed more arch and spring of rib, was deeper flanked, and heavier muscled as exhibited by a thicker top and wider, squarer dock. 4 also appeared to have a higher weight per day of age.

I admit that 3 was a cleaner-fronted ewe, but she simply lacked the performance, volume, and muscle to place any higher. Thank you.

Rams

Besides the sheep terms already given, the following terms can be used to describe classes of rams (also known as **bucks**).

DESIRABLE	UNDESIRABLE
Greater scrotal circumference	Less scrotal circumference
More powerful	
More masculine	
More ruggedly designed	
Stouter made	

DESIRABLE	UNDESIRABLE
More apt to sire lambs with more:	
frame	
growth	
trimness	
performance	

INCORPORATION OF PERFORMANCE DATA INTO REASONS

If performance data are provided for a class of breeding livestock, they should be studied intently prior to evaluating the class. In fact, many successful judges place the performance data before analyzing the livestock, and then make no more than pair switches for their final placing. When data are provided, they should be worked into a set of reasons using statements to emphasize places where the data analysis played a large role in the placing. Reasons emphasis should be placed where the data match the livestock as well as where they do not.

SUMMARY

Market livestock are evaluated on their ability to produce heavily-muscled, lean carcasses. Structural correctness, performance, and muscle pattern must also be assessed. Evaluation of breeding livestock relies on similar parameters, but structure, volume, performance, and breeding soundness play a much more important role.

Terminology used for reasons is also similar between market and breeding livestock within the same species. However, terminology differs among the livestock species. Livestock judges must memorize the parts of market and breeding livestock. Care must be taken not to mix terminology among species.

Becoming a good livestock judge takes work and practice. Only a limited amount of livestock judging can be learned from a book. The rest must be learned from observing livestock and evaluating the strengths and weaknesses of individual animals.

CHAPTER REVIEW

EXPERIENTIAL LEARNING

1. Livestock judging helps students make selections based on industry and individual producer needs. Consider forming a livestock judging team and entering contests at the local and state level.

2. Students with livestock judging projects can arrange classes for a county or regional livestock judging practice session. Invite an experienced evaluator to help with placings and reasons.

DEFINE ANY TEN KEY TERMS

blind nipples

bowlegged (front legs)

bowlegged (rear legs)

buck-kneed

bucks

calf-kneed

cowhocked

finish

flat nipples

knock-kneed

pigeon-toed

pin nipples

posty

sickle-hocked

splay-footed

QUESTIONS AND PROBLEMS FOR DISCUSSION

1. Name the three judging parameters of selecting market livestock.

2. True or False. Market livestock are most often evaluated using performance data.

3. Highest-priced cuts of meat are located in the front/rear half of the animal. Select one.

4. Finish is another term for:

   a. muscle

   b. bone

   c. age

   d. fat cover

5. Explain where fat cover can be observed in a beef animal.

6. List two terms describing undesirable fat cover in beef.

7. What does posty mean?

8. True or False. Legs that bow out are called bowlegged.

9. Name three places to look for muscling on a market hog.

10. When viewed from behind, would a hog with a square or rounded top be fatter?

11. How do judging contestants usually determine muscling in lambs?

12. In which species would you use the terminology heavier-muscled hindsaddle?

13. Give the ideal fat cover for lambs in inches.

14. Describe the front end of a feminine heifer.

15. True or False. Scrotal circumference indicates sperm production potential.

16. A gilt should have at least _____ nipples per side.

17. _____ nipples are not functional.

18. Differentiate between ewe lambs and yearlings.

19. List another term for ram.

20. Which is more desirable, a dense or open fleece?

Objectives

After completing this chapter, students should be able to:

▶ Score a cow using the Dairy Cow Unified Score Card

▶ Differentiate between type and linear classification

▶ Describe the desirable traits of an ideal dairy cow

▶ Judge and give reasons on a class of dairy cows

Key Terms

colored breeds

Dairy Cow Unified Score Card

lateral suspensory ligament

linear classification

median suspensory ligament

predicted transmitting ability for type (PTAT)

type classification

Career Focus

Bill grew up on a small Northeast dairy. He showed Guernsey heifers while in FFA, majored in dairy science in college, and participated in the college dairy judging team. His dream job was to be a classifier for a national dairy breed association but knew those positions did not open very often. So to further prepare for a career as a classifier, Bill worked for several years as an artificial insemination (AI) technician for a large dairy genetics company. The company had an opening for a person to search the country for exceptional cows that might serve as potential dams for young AI sires. Bill applied for and landed the job. During his forays around the United States in search of high-quality dairy cows, Bill met many top breeders and influential people in the world of dairy genetics. He gained reputation as an exceptional evaluator of cow type and even judged some national dairy shows. Finally, an opening arrived for a classifier. Bill's resume was well stocked with experience and his letters of reference were excellent. Bill hopes to spend the rest of his career classifying cows.

INTRODUCTION

Selection of dairy cattle should be primarily based on the cow's ability to produce large quantities of milk for several years. Thus, the milk factory, or udder is extremely important when evaluating dairy cattle

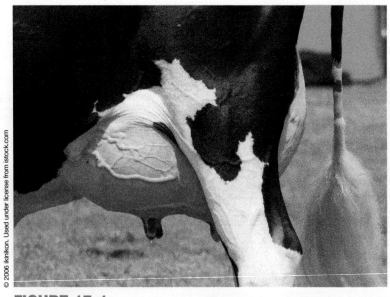

© 2006 ikinikon. Used under license from istock.com

FIGURE 17-1

Dairy cow judging centers around the mammary gland.

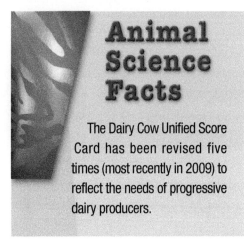

Animal Science Facts

The Dairy Cow Unified Score Card has been revised five times (most recently in 2009) to reflect the needs of progressive dairy producers.

type. The rest of the cow can be viewed as an udder support system. Feet and legs must be correct so the udder can move about easily for the productive life of the animal. In addition, feet and legs must be constructed properly so the cow can move around efficiently and comfortably to consume large quantities of feed and water. The body should be wide and deep to hold feed that the udder eventually converts to milk. In plain terms, dairy cattle judging centers around the mammary gland.

TYPE CLASSIFICATION

As with any livestock species, knowledge of the external parts precedes further critical discussion.

Dairy cows can be evaluated by comparing each cow to an ideal cow, then assigning a score. The Purebred Dairy Cattle Association made this comparison easier by establishing the **Dairy Cow Unified Score Card** in 1943. See figure 17-2 that shows parts of the cow. The current score card divides the dairy cow into four major traits, each consisting of several sub-elements. Each sub-element is assigned a specific point value. Sub-element point values are evaluated and scored to arrive at a final score for the trait. Total point values for the four traits total 100 (frame— 15 points, dairy strength—25 points, rear feet and legs—20 points, and udder—40 points). Traits and the corresponding point values are adjusted periodically to reflect changes in ideal cow type for current management systems.

Breed associations have trained classifiers who visit farms and classify registered cattle. Presently, the Holstein Association provides this service for some of the smaller **colored breeds** (breeds other than Holsteins). Using the Dairy Cow Unified Score Card, classifiers rate registered cattle. Classified cows fit into type categories according to their individual score: 90 to 100 excellent, 85 to 89 very good, 80 to 84 good plus, 75 to 80 good, 70 to 74 fair, less than 70 poor. See figure 17-3. This process is known as **type classification**. Herds receive a summary score indicating the overall rating of the herd. Classification assists dairy producers with

DAIRY COW UNIFIED SCORECARD

Breed characteristics should be considered in the application of this scorecard.

MAJOR TRAIT DESCRIPTIONS

Perfect Score

There are four major breakdowns on which to base a cow's evaluation. Each trait is broken down into body parts to be considered and ranked.

1) Frame - 15%

15

The skeletal parts of the cow, with the exception of rear feet and legs. Listed in priority order, the descriptions of the traits to be considered are as follows:

Rump (5 points): Should be long and wide throughout. Pin bones should be slightly lower than hip bones with adequate width between the pins. Thurls should be wide apart. Vulva should be nearly vertical and the anus should not be recessed. Tail head should set slightly above and neatly between pin bones with freedom from coarseness. **Front End (5 points):** Adequate constitution with front legs straight, wide apart, and squarely placed. Shoulder blades and elbows set firmly against the chest wall. The crops should have adequate fullness blending into the shoulders. **Back/Loin (2 points):** Back should be straight and strong, with loin broad, strong, and nearly level. **Stature (2 points):** Height including length in the leg bones with a long bone pattern throughout the body structure. Height at withers and hips should be relatively proportionate. Age and breed stature recommendations are to be considered. **Breed Characteristics (1 point):** Exhibiting overall style and balance. Head should be feminine, clean-cut, slightly dished with broad muzzle, large open nostrils and strong jaw.

2) Dairy Strength - 25%

25

A combination of dairyness and strength that supports sustained production and longevity. Major consideration is given to general openness and angularity while maintaining strength, width of chest, spring of fore rib, and substance of bone without coarseness. Body condition should be appropriate for stage of lactation. Listed in priority order, the descriptions of the traits to be considered are as follows:

Ribs (8 points): Wide apart. Rib bones wide, flat, deep, and slanted towards the rear. Well sprung, expressing fullness and extending outside the point of elbows. **Chest (6 points):** Deep and wide floor showing capacity for vital organs, with well-sprung fore ribs. **Thighs (2 points):** Lean, incurving to flat and wide apart from the rear. **Neck (2 points):** Long, lean, and blending smoothly into shoulders; clean-cut throat, dewlap, and brisket. **Withers (2 points):** Sharp with chine prominent. **Skin (1 point):** Thin, loose, and pliable.

3) Rear Feet and Legs - 20%

20

Feet and rear legs are evaluated. Evidence of mobility is given major consideration. Listed in priority order, the descriptions of the traits to be considered are as follows:

Movement (5 points): The use of feet and rear legs, including length and direction of step. When walking naturally, the stride should be long and fluid with the rear feet nearly replacing the front feet. **Rear Legs-Side View (3 points):** Moderate set (angle) to the hock. **Rear Legs-Rear View (3 points):** Straight, wide apart with feet squarely placed. **Feet (3 points):** Steep angle and deep heel with short, well-rounded closed toes. **Thurl Position (2 points):** Near central placement between the hip and pin bones. **Hocks (2 points):** Adequate flexibility with freedom from swelling. **Bone (1 point):** Flat and clean with adequate substance. **Pasterns (1 point):** Short and strong with some flexibility, having a moderate, upright angle.

4) Udder - 40%

40

The udder traits are evaluated. Major consideration is given to the traits that contribute to high milk yield and a long productive life. Listed in priority order, the descriptions of the traits to be considered are as follows:

Udder Depth (10 points): Moderate depth relative to the hock with adequate capacity and clearance. Consideration is given to lactation number and age. **Rear Udder (9 points):** Wide and high, firmly attached with uniform width from top to bottom and slightly rounded to udder floor. **Teat Placement (5 points):** Squarely placed under each quarter, plumb and properly spaced. **Udder Cleft (5 points):** Evidence of a strong suspensory ligament indicated by clearly defined halving. **Fore Udder (5 points):** Firmly attached with moderate length and ample capacity. **Teats (3 points):** Cylindrical shape; uniform size with medium length and diameter; neither short nor long is desirable. **Udder Balance and Texture (3 points):** Udder floor level as viewed from the side. Quarters evenly balanced; soft, pliable, and well collapsed after milking. **(Note: In the Holstein breed, an equal emphasis is placed on fore and rear udder (7 points each). All other traits are the same as listed above.)**

TOTAL

100

Copyrighted by the Purebred Dairy Cattle Association, 1943. Revised and copyrighted 1957, 1971, 1982, 1994 and 2009.

FIGURE 17-2

Critical information for dairy cattle judges resides on the Dairy Cow Unified Score Card.

THE SEVEN BREEDS

Ayrshire   Brown Swiss   Guernsey

Red & White   Holstein   Jersey   Milking Shorthorn

BREED CHARACTERISTICS

Except for differences in color, size, and head character, all breeds are judged on the same standards as outlined in the Unified Score Card. If any animal is registered by one of the dairy breed associations, no discrimination against color or color pattern is to be made.

Ayrshire   Strong and robust, showing constitution and vigor, symmetry, style and balance throughout, and characterized by strongly attached, evenly balanced, well-shaped udder.
HEAD- clean cut, proportionate to body; broad muzzle with large, open nostrils; strong jaw; large, bright eyes; forehead, broad and moderately dished; bridge of nose straight; ears medium size and alertly carried.
COLOR- light to deep cherry red, mahogany, brown, or a combination of any of these colors with white, or white alone, distinctive red and white markings preferred.
SIZE- a mature cow in milk should weigh at least 1200 lbs.

Brown Swiss   Strong and vigorous, but not coarse. Adequate size with dairy quality. Frailness undesirable.
HEAD- clean cut, proportionate to body; broad muzzle with large, open nostrils; strong jaw; large, bright eyes; forehead, broad and slightly dished; bridge of nose straight; ears medium size and alertly carried.
COLOR- body and switch solid brown varying from very light to dark; muzzle has black nose encircled by a white ring; tongue and hooves are dark brown to black.
SIZE- a mature cow in milk should weigh at least 1400 lbs.

Guernsey   Strenth and balance, with quality and character desired.
HEAD- clean cut, proportionate to body; broad muzzle with large, open nostrils; strong jaw; large, bright eyes; forehead, broad and slightly dished; bridge of nose straight; ears medium size and alertly carried.
COLOR- shade of fawn and white markings throughout clearly defined.
SIZE- a mature cow in milk should weigh 1200-1300 lbs.; Guernsey does not discriminate for lack of size.

Red & White   Rugged, feminine qualities in an alert cow possessing adequate size and vigor.
HEAD- clean cut, proportionate to body; broad muzzle with large, open nostrils; strong jaw; large, bright eyes; forehead, broad and slightly dished; bridge of nose straight; ears medium size and alertly carried.
COLOR- must be clearly defined red and white; black-red and brindle is strictly prohibited.
SIZE- a mature cow in milk should weigh at least 1400 lbs. and be well balanced.

Holstein   Rugged, feminine qualities in an alert cow possessing Holstein size and vigor.
HEAD- clean cut, proportionate to body; broad muzzle with large, open nostrils; strong jaw; large, bright eyes; forehead, broad and moderately dished; bridge of nose straight; ears medium size and alertly carried.
COLOR- black and white or red and white markings clearly defined
SIZE- a mature cow in milk should weigh at least 1400 lbs.
UDDER- equal emphasis is placed on fore and rear udder (7 points each), all other traits are the same as listed on the PDCA scorecard.

Jersey   Sharpness with strength indicating productive efficiency.
HEAD- proportionate to stature showing refinement and well chiseled bone structure. Face slightly dished with dark eyes that are well set.
COLOR- some shade of fawn with or without white markings; muzzle is black encircled by a light colored ring; switch may be either black or white.
SIZE- a mature cow in milk should weigh at least 1000 lbs.

Milking Shorthorn   Strong and vigorous, but not coarse.
HEAD- clean cut, proportionate to body; broad muzzle with large, open nostrils; strong jaw; large, bright eyes; forehead, broad and slightly dished; bridge of nose straight; ears medium size and alertly carried.
COLOR- red or white or any combination (no black markings allowed).
SIZE- a mature cow in milk should weigh 1400 lbs.

FACTORS TO BE EVALUATED

The degree of discrimination assigned to each defect is related to its function and heredity. The evaluation of the defect shall be determined by the breeder, the classifier or judge, based on the guide for discrimination and disqualifications given below.

HORNS
No discrimination for horns.
EYES
1. Blindness in one eye: *Slight discrimination.*
2. Cross or bulging eyes: *Slight discrimination.*
3. Evidence of blindness: *Slight to serious discrimination.*
4. Total blindness: *Disqualification.*
WRY FACE
Slight to serious discrimination.
CROPPED EARS
Slight discrimination.
PARROT JAW
Slight to serious discrimination.
SHOULDERS
Winged: Slight to serious discrimination.
CAPPED HIP
No discrimination unless affects mobility.

TAIL SETTING
Wry tail or other abnormal tail settings: *Slight to serious discrimination.*
LEGS AND FEET
1. Lameness- apparently permanent and interfering with normal function: *Disqualification.* Lameness- apparently temporary and not affecting normal function: *Slight discrimination.*
2. Evidence of crampy hind legs: *Serious discrimination.*
3. Evidence of fluid in hocks: *Slight discrimination.*
4. Weak pastern: *Slight to serious discrimination.*
5. Toe out: *Slight discrimination.*
UDDER
1. Lack of defined halving: *Slight to serious discrimination.*
2. Udder definitely broken away in attachment: *Serious discrimination.*
3. A weak udder attachment: *Slight to serious discrimination.*
4. Blind quarter: *Disqualification.*
5. One or more light quarters, hard spots in udder, obstruction in teat (spider): *Slight to serious discrimination.*

6. Side leak: *Slight discrimination.*
7. Abnormal milk (bloody, clotted, watery): *Possible discrimination.*
LACK OF ADEQUATE SIZE
Slight to serious discrimination. (Note: Guernsey does not discriminate for lack of size.)
EVIDENCE OF SHARP PRACTICE
(Refer to PDCA Code of Ethics)
1. Animals showing signs of having been tampered with to conceal faults in conformation and to misrepresent the animal's soundness: *Disqualification.*
2. Uncalved heifers showing evidence of having been milked: *Slight to serious discrimination.*
TEMPORARY OR MINOR INJURIES
Blemishes or injuries of a temporary character not affecting animal's usefulness: *Slight to serious discrimination.*
OVERCONDITIONED
Slight to serious discrimination.
FREEMARTIN HEIFERS
Disqualification.

FIGURE 17-2 *(continued)*

Photo courtesy of Delmar and Larry Zimmerman

FIGURE 17-3

This Holstein, named Walnut-Hills Logic Cathy, scored 95 classification points, or Excellent status.

merchandising and selection of breeding replacements and sires. An objective opinion of cattle allows better decision making on the part of producers.

The first trait in the Dairy Cow Unified Score Card, frame, includes all skeletal parts of the cow except for the rear legs. Frame can receive up to 15 points. The sub-elements of rump (5 points), front end (5 points), back and loin (2 points), stature (2 points), and breed characteristics (1 point) make up this trait.

The second trait, dairy strength, assesses the physical characteristics that should allow the cow to sustain production for many years. Ribs (8 points), chest (6 points), barrel (4 points), thighs (2 points), neck (2 points), withers (2 points), and skin (1 point) are all elements used to score dairy strength. A perfect dairy strength score would earn 25 points. See figure 17-4.

The third trait, rear feet and legs can count for up to 20 points. Movement (5 points), rear legs–side view (3 points), rear legs–rear view (3 points), feet (3 points), thurl position (2 points), hocks (2 points), bone (1 point), and pasterns (1 point) determine the points scored in this category. See figure 17-5.

The most important of the four traits, the udder is analyzed from all angles. Udder depth (10 points), rear udder (9 points), teat placement (5 points), udder cleft (5 points), fore udder (5 points), teats (3 points),

Photo courtesy of Delmar and Larry Zimmerman

FIGURE 17-4

Dairy strength helped this cow, Walnut-Hills Astro Brianne, to 94 classification points.

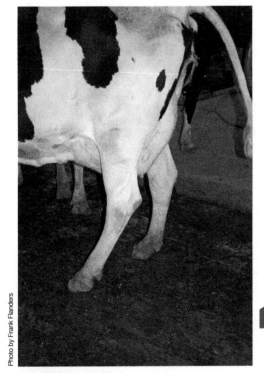

Photo by Frank Flanders

FIGURE 17-5

This cow has too much set to her rear leg.

as well as udder balance and texture (3 points), tally up to 40 points in this category. In Holsteins, the weighting of points for the fore and rear udder is slightly altered. Seven points are assessed to each of these for Holsteins only.

Page 1 of the Dairy Cow Unified Score Card gives excellent descriptions of the ideal elements for each trait. Study and memorize the elements.

Page 2 of the Dairy Cow Unified Score Card provides pictures of cows from all seven of the major dairy breeds. Use these pictures as examples of ideal dairy cows of each breed. Ideal breed characteristics are listed beneath the pictures. At the bottom of the card, a guide of defects and conditions, along with degrees of discrimination, is given. Use the guide to assess the importance of any physical flaws you may encounter.

LINEAR CLASSIFICATION

Linear classification carries dairy type traits into the realm of genetics. Linear classification gives dairy producers a method to utilize the heritability of individual type traits. A scale from −3 to +3 is used to

Photo courtesy of Penn State Department of Dairy and Animal Science

FIGURE 17-6

Penn State Ivanhoe Star was one of the famous types and production bulls of his time.

score the linear traits. Teat placement, teat width, rear legs from side, rear legs from rear, and foot angle are some traits considered in linear evaluation. Producers can select a potential sire by evaluating his score for an individual trait and then matching his stronger type traits with a dam's weaknesses. These linear trait scores are combined to calculate **predicted transmitting ability for type (PTAT)**, an overall type indicator. Most PTAT scores usually range from 0.5 to 2.5. A 2.18 PTAT bull would be a better type sire selection than a bull with a PTAT of 0.5. Resulting positive changes in structural or udder types from selection based on linear evaluation can add years to the life of the average cow in a given herd. See figure 17-6.

JUDGING DAIRY COWS

Classes of dairy cows are evaluated much the same as individual dairy cows. Completing a Dairy Cow Unified Score Card for each cow and then comparing the final scores is the simplest way to place classes of dairy cows. Reasons should evolve from the differences found in the elements of the scorecard.

Terminology used in reasons for dairy cattle originates from the Dairy Cow Unified Score Card as well
as the parts of the dairy cow. Positive comparison terms
should be used, while negative comparisons should be
avoided when possible. An example of a positive term
would be to refer to a cow as longer bodied. The comparative negative term would be to refer to a cow as
shorter bodied. A breakdown of the scorecard traits,
with terms used to describe the elements, follows.

FRAME

The ideal dairy cow is large framed, long bodied, and
correctly structured. Hooks and pins are wide, and
the rump is nearly level, with preference given for the
pins to be slightly lower than the hooks, which aids in
the birth of the calf. The shoulder, neck, and head are
clean, feminine, and angular. See figure 17-7.

General Terms

More style, balance, and symmetry

Smoother blending throughout

Taller, longer, more upstanding

More harmonious blending of body parts

More strength and power

More angular

Milkier

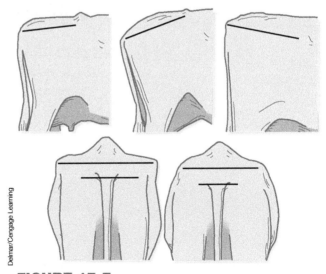

Delmar/Cengage Learning

FIGURE 17-7

Differences in rump slope and width of hooks and pins.

Size and Scale

Taller

More size and scale

More upstanding

Longer bodied

Framier

Head

More alert

Broader through the muzzle

More feminine about the head

Longer, more balanced head

Cleaner, more refined head

Front end (Shoulders)

Smoother through the front end

Tighter and smoother at the point of the shoulder

More neatly laid in at the shoulders

Blends more smoothly from neck to shoulders

Tighter and fuller at the point of the elbow

Shoulders blend more smoothly into the body wall

Sharper over the withers

Fuller at the crops

Topline and rump

Straighter over the topline

Cleaner over the topline

Harder down the topline

Stronger in the chine

Stronger and harder in the loin

More nearly level from hooks to pins

Smoother tailhead

Neater tail setting

Wider through the (hooks, pins, thurls, rump)

More level through the chine and loin

More refined tailhead

Higher chine

More prominent about the hooks and pins

Longer from hooks to pins

Cleaner over the rump

Front legs

Stands straighter on her front legs

Walks out more correctly on her front feet

DAIRY STRENGTH

Dairy strength describes the combination of capacity and femininity that lend a cow to a long, productive milking life. The volume of nutrients that a cow takes ultimately affects the amount of milk produced. A cow must have space to hold the nutrients (feed) she eats. Body capacity is evaluated by the width and depth of ribs, and the size of the barrel. Cows with good dairy strength are deep and wide, yet long fronted and feminine. Body condition should be appropriate for the stage of lactation, neither too fat nor too thin. See figures 17-8 and 17-9.

General Capacity Terms

Wider chest

Deeper chest

Deeper barrel

Displays more spring of rib

© WilleeCole. Used under license from shutterstock.com

FIGURE 17-8

This Jersey carries too much body condition to score well in dairy character.

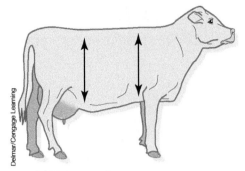 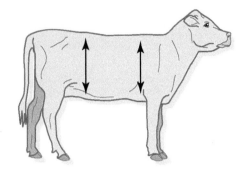

FIGURE 17-9

The cow on the left is much deeper in her rib.

Deeper in the flank

More overall depth and capacity

More depth of heart

Deeper in the fore rib

Deeper in the rear rib

More depth of barrel

More depth and width of body

General Femininity Terms

Sharper, cleaner cut, more angular throughout

Excels in dairy character / strength

Cleaner and more feminine throughout

Neck

Longer, leaner neck

More clean-cut head and neck

Trimmer neck, throat, dewlap, and brisket

More refined through the head and neck

Thighs

More incurving in the thighs

Cleaner, flatter, and thinner in the thighs

Body

Sharper over the withers

Cleaner over the topline

Displays more prominent vertebrae

More prominent hooks and pins

Cleaner rump

Flatter and cleaner bone

More open rib

REAR FEET AND LEGS

Common flaws in feet and legs are much the same as those in beef cattle. A correctly set rear legs and feet are vital to a cow's longevity. Cows with a correct pastern angle and a deep hoof heel wear the hoof evenly, decreasing the need for costly hoof trimming. See figure 17-10.

Rear Leg Stance

Straighter legs from side view (rear view)

Stands more squarely on her rear legs

Shows more moderate set to her hock

Has stronger pasterns

Cleaner in the hocks

Feet

Has a deeper heel

Has a more correctly shaped hoof

Stands on a shorter, more shapely hoof

Movement

Tracks straighter on her rear legs

Moves more gracefully

Moves more freely in/through the hock

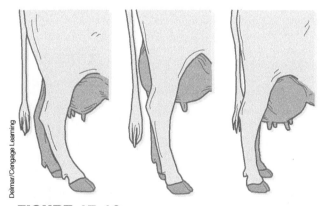

Delmar/Cengage Learning

FIGURE 17-10

The cow in the center picture has the correct set to her rear leg.

UDDER

As mentioned previously, the udder is the most important single trait upon which dairy cows are judged. In a dairy cattle judging contest, the cow with the best udder wins the class most of the time. Therefore, the ability of judges to analyze udders and identify differences among udders becomes critical to success.

Ideally, an udder should be deep to provide capacity for milk production. However, the udder should not drop below the hock. Udders that drop too far are more susceptible to bacterial invasions and are more prone to physical damage, such as teat tramping. Young cows will have higher, tighter udders than older cows. However, youthfulness of the mammary system should always be considered in relation to the age of an animal. See figure 17-11.

Teats should be placed squarely beneath each quarter and should hang perpendicular to the ground. The distance between the front teats and the distance between the rear teats should be the same. Teats should be of uniform size, intermediate length, and moderate diameter. See figure 17-12.

The rear of the udder should be attached high when viewing the cow from the rear. A high rear udder attachment signifies added udder volume. The width of the udder should be uniform from top to bottom. The **median suspensory ligament**, which divides the udder in half from side to side, should be plainly visible. See figure 17-13. The **lateral suspensory ligament** separates the rear and fore udder. See figure 17-14. Cows without obvious ligaments tend to have udders that drop too low, leaving the udder prone to injury. The base of the udder should

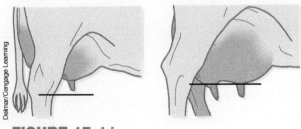

FIGURE 17-11

The udder should not fall below the hock.

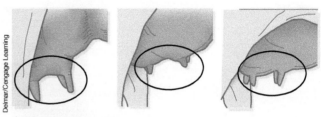

FIGURE 17-12

Ideal teat size, as displayed by the cow on the right, makes milking easier.

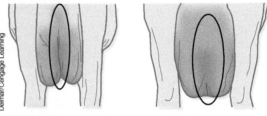

FIGURE 17-13

The median suspensory ligament creates udder cleft when viewed from the rear. Cows with an obvious suspensory ligament typically last longer in the herd before udder problems force culling.

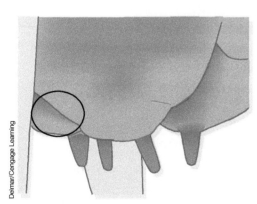

FIGURE 17-14

The cleft separating the fore udder from the rear udder is caused by the lateral suspensory ligament.

be relatively level without a funneling effect between the fore and rear quarters. Alternately, a weak or broken suspensory ligament can cause the teats to jut or strut outward.

The fore udder should blend smoothly into the body wall and show a strong attachment to the body wall. See figure 17-15. In addition, the fore udder should be deep and moderately long but not disproportionate to the rest of the udder. All quarters should be evenly balanced, and the udder floor should be level when viewed from the side. See figure 17-16. Moreover, udder veining is an indication of the amount of blood passing through the udder, so veining should be prominent. See figure 17-17.

General Terms

More shapely, symmetrical udder

More balanced udder

More level udder floor

Udder carried higher above the hock

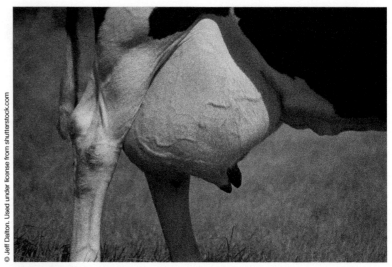

FIGURE 17-15

The udder on this cow does not blend smoothly into the body wall.

FIGURE 17-16

The udder should be nearly level, not sloped from front to back like this cow.

More capacious udder

More youthful mammary system

More apparent quality and texture of udder

Fore Udder

Longer fore udder attachment

Smoother blending fore udder

Blends more smoothly into the body wall

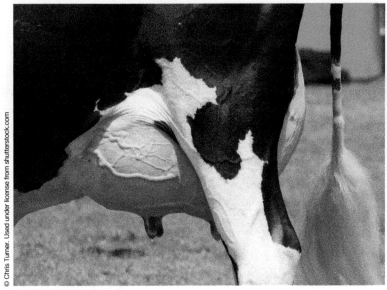

FIGURE 17-17
Note the prominent veining on this udder signifying tremendous blood flow.

Rear Udder

Higher, wider rear udder

More uniform width of rear udder

More defined halving

Stronger median suspensory ligament

Fuller rear udder

Teats

Closer front teat placement

Front teats more centrally placed

Teats hanging more nearly plumb

Teats more evenly spaced beneath quarters

More desirable teat size

More desirable teat shape

Shorter teats

Teats more neatly set on the udder floor

SAMPLE REASONS

Holstein four-year-old cows, Placing: 3–2–1–4

In these 4-year old Holstein, 3 and 2 show a definite advantage in dairy strength over 1 and 4. I placed

Animal Science Facts

FFA members can potentially compete in two dairy career development events at the National FFA Convention: dairy food judging and dairy cattle judging. Other FFA members act as handlers during the dairy cattle judging contest.

3 first and over 2 because of her advantage in mammary system. 3 displayed a higher, wider rear udder and a more defined suspensory ligament, as viewed from the rear. Furthermore, 3's udder is more nearly level when viewed from the side and has shorter teats. I grant that 2 is sharper in her chine and more incurving in her thighs.

In my middle pair, 2 easily placed over 1 on her overall general appearance, strength, and power. 2 is taller in her front end and harder down her topline. 2 is also tighter in the shoulder, longer and cleaner in her neck, and more refined about the head. She is deeper in the forerib and barrel, and more open in her rib with cleaner, flatter bone. In her mammary, 2 is higher and wider in her rear udder. I concede that 1 is deeper in the heel.

In my final placing, 1 edges the white cow as she displays a more youthful mammary system by carrying her udder higher above the hocks, having a longer, smoother fore udder, and a stronger suspensory ligament. 1 also is more refined and angular about her front end.

Although 4 is a longer-bodied cow, I placed 4 last in this class for being too deep in the udder and coarse throughout.

For these reasons, I placed this class of Holstein 4-year-old cows 3–2–1–4.

JUDGING DAIRY HEIFERS

Heifers are evaluated on the same general traits as cows in production with the notable exception of the udder. Stature, style, dairy strength, and structural correctness are the factors upon which placings should be made. Terms are the same as those used for cows in production. See figure 17-18 for a picture of a very good dairy heifer.

SAMPLE REASONS

Guernsey Intermediate Heifers, Placing: 1–3–4–2

In my top placing, 1 places over 3 on her obvious advantage in body capacity. 1 exhibits more spring

Photo courtesy of Delmar and Larry Zimmerman

FIGURE 17-18

A former Reserve Junior All-American Summer Yearling Holstein heifer.

of rib and is deeper in both heart and barrel. Also, 1 is wider in the chest floor, wider at the pins, and neater at her tailhead. I concede 3 is longer, cleaner necked, and shows more breed character about the head.

In my middle pair, 3 follows 1's pattern of style and straightness of lines to place over 4. 3 is taller fronted, longer bodied, harder and higher in her chine, and lower at her pins. Furthermore, 3 is cleaner and more angular about her front end, being especially cleaner about the dewlap and smoother at the point of her shoulder. I recognize 4 is deeper in her fore rib and tracks straighter on her rear legs.

In my final placing, 4 logically places over 2 with her advantage in dairy strength and feet and legs. 4 is a cleaner, flatter-boned heifer with more open ribbing and more incurving thighs. 4 stands straighter on her rear legs, when viewed from the side, has stronger pasterns, and is deeper in her heel. I grant that 2 is wider at her hooks.

Although 2 is a long-bodied, neat-fronted heifer, she lacks the strength, power, body depth, and correctness of hock found in the other three heifers.

For these reasons, I placed this class of Guernsey Intermediate Heifers 1–3–4–2.

SUMMARY

Dairy cattle are judged with an emphasis on the traits needed to produce milk and to remain productive over several years. The Dairy Cow Unified Score Card lists and assigns relative weighting of these traits. The udder category receives the highest point value. Linear classification is used to predict the ability to genetically alter type traits. Dairy terms are distinctly different from those used for livestock judging classes. A working vocabulary of the parts of the dairy cow is essential to delivering good reasons. Except for the udder, heifers are judged on the same traits as cows in production.

CHAPTER REVIEW

EXPERIENTIAL LEARNING OPPORTUNITIES

1. Students who do not live or work on a dairy can still be successful dairy judges. Students can learn about dairy judging opportunities by visiting www.FFA.org or contacting an extension educator.

2. Job shadow experiences are plentiful in the dairy industry. Students can seek opportunities with AI technicians to develop an appreciation for sire selection.

DEFINE ALL TERMS

colored breeds
Dairy Cow Unified Score Card
lateral suspensory ligament
linear classification

median suspensory ligament
predicted transmitting ability for
   type (PTAT)
type classification

QUESTIONS AND PROBLEMS FOR DISCUSSION

1. True or False. AI stands for artificial insemination.

2. When judging dairy cows using the Dairy Cow Unified Score Card, the _____ receives the most consideration.

3. List the four major traits found on the Dairy Cow Unified Score Card.

4. Write perfect scores for the Dairy Cow Unified Score Card traits.

5. True or False. A type score of 70 to 79 would classify a cow as "good."

6. Type classification conducted by a trained classifier rates registered dairy cows on scale worth up to _____ points.

7. List two defects that can lead to disqualification during classifying.

8. Teats should be placed squarely beneath each _____ and hang _____ to the ground.

9. Differentiate between a cow's heart girth and barrel.

10. True or False. Fat dairy cows are the best milk producers.

11. Which is located closer to a cow's head, the hooks or the pins?

12. Dairy strength includes:

    a. ribs

    b. chest

    c. rear leg structure

    d. both a and b

13. Linear classification rates individual traits on a scale of _____
    to _____.

14. Linear traits are combined to calculate _____.

15. Why should dairy cows not carry an abundance of condition or flesh?

16. Jutting or strutting teats indicate a weak _____.

17. Why does the udder not receive primary attention when judging heifers?

18. List two career skills contests in which FFA members interested in dairy can participate.

19. The Dairy Cow Unified Score Card has been revised most recently in _____.

20. True or False. Terminology used in dairy judging is the same as that used for beef judging.

GLOSSARY

A

abomasum—True stomach in ruminants

actin—Thin muscle protein

adipocytes—Individual fat cells

adipose tissue—Fat tissue

aerobic exercise—Long-term, less intense exercise

AHC—American Horse Council

all-in all-out production—Swine management where all pigs are removed from a building or room before the next group of pigs arrives

alleles—Pairs of genes

amino acid—Building block for proteins

anaerobic exercise—Intense, short-term exercise

androgens—Male sex hormones

animal rightists—Believe animals should have the same rights and privileges as humans

animal welfarists—Believe animals should be healthy, happy, and free from want

anthelmintic—Worming medicine

apoptosis—Programmed cell death

atrophic rhinitis—Disease causing degeneration of turbinate bones in the snout of swine

average daily gain (ADG)—Measures rate of gain

B

backcrossing—Two-breed crossbreeding system

backfat at 250 pounds (BF)—Subcutaneous fat, adjusted to a 250 pound equivalent in swine

backgrounding—Placing cattle on low-energy, high-forage, low-cost diet for their first winter and then in a feedlot the following spring

baldies—White faced cattle

band—Herd of mares

barrows—Male swine with testicles removed

beta agonist—Synthetic substances similar to adrenaline that improve the lean:fat ratio

binomial nomenclature—System of naming animals that includes only the genus and species

biosecurity—Methods to reduce disease transfer

birth weight EPD (BW EPD)—Expected progeny difference for birth weight in beef

black baldies—Hereford x Angus crossbreds

blind nipples—Nonfunctional nipples in swine

boars—Intact male swine

body condition—Muscle and fat cover

bowlegged (front legs)—Front legs bow out (front view)

bowlegged (rear legs)—Rear legs bow out (rear view)

break joint—Red porous front leg bone of a lamb carcass indicating a lamb younger than 1 year.

brood cow—Mother beef cows

buck-kneed—Front legs too straight or knees forward of the hoof (side view)

bucks—Another name for intact male sheep

bulls—Intact male cattle

buttons—Bits of cartilage along split surface of a beef carcass

C

calf-kneed—Front legs too curved (side view)

capacitation—Process where fluids present in the female reproductive tract wash the sperm and change its chemical composition making it capable of fertilization

cartilage—Precursor to hard bones

cash price—Live price paid by a packer on a given day

castrated—Testicles removed

cecum—An organ at the beginning of colon of horses that supports a population of digestive bacteria; appendix in humans

cell—Smallest biological unit

cell membrane—Outer membrane of an animal cell

centromere—Attachment point for chromatids

cervix—Seals off uterus after pregnancy

check-off—Money collected from every animal sold used for product promotion

choice/select spread—Difference between carcass price of most common beef quality grades

chromosomes—Condensed strands of genetic material in the nucleus

chyme—Partially digested material

clip—Fleeces from the sheep flock

codominance—Blending of alleles

colic—Abdominal pain

collagen—Fibrous tissue stronger than steel

colored breeds—Dairy breeds other than Holsteins

colostrum—First milk

commercial cattle—Crossbred beef animals for commercial production

compensatory gain—Profitable weight gain in thin cattle made at the start of the finishing period

complete dominance—One gene completely masks the other's presence

complex carbohydrates—Carbohydrates that cannot be digested by monogastric animals, for example, cellulose and hemicellulose

conception—Time of fertilization

conditioning—Association of certain stimulus with a behavior

contemporary groups—Group of animals fed and managed in the same way

continental breed—Beef breed from the European continent

cooperatives—Business with share in ownership

corpus hemorrhagicum—Bloody body on the ovary that held the egg prior to ovulation

corpus luteum—Yellow body on the ovary that produces a hormone called progesterone, which sustains pregnancy

cow-calf—Beef operation maintaining cows and calves

cowhocked—Rear legs bow in (rear view)

creep feeding—Allowing young nursing animals access to high-quality feed

crop—Structure located at the base of the neck in poultry that serves as a storage area for recently ingested feed

cross-fostered—Piglets transferred to another sow that produces more milk

crossbreeding—Mating of animals that are not related; also known as out breeding

crutched—Shearing dirty wool around vulva and udder

cull—Eliminate or sell

cut—Indicates the difficulty of a placing in a judging contest

cutability—Percent lean of a carcass

cwt.—100 pounds

cytoplasm—Jelly-like substance in cells between the cell membrane and the nucleus

D

Dairy Cow Unified Score Card—Compares a dairy cow to the ideal cow

days to 250 pounds (DAYS)—Measures lifetime growth to 250 pounds in swine

DHIA (Dairy Herd Improvement Association)—Record-keeping organization for dairy animals

displaced abomasums—Twisted stomach; occurs when the abomasum moves to an abnormal position in the body cavity

distemper (horses)—Streptococcal infection in young horses; also called strangles

DNA—Deoxyribonucleic acid, the genetic compound that controls inheritance

docked—Tail removed

draft horses—Heavy work horses

drench—Place medicine in back of the throat

dressage—Competition where riders guide horses through unique maneuvers

dressing percentage—Carcass percent of live weight

dry cows—Nonlactating female cattle

dystocia—Calving difficulty

E

eared cattle—Brahman cattle

efferent ducts—Larger sperm collection area after rete testis

eliminative behaviors—Defecating behaviors

encephalomyelitis—Sleeping sickness in horses

endomysium—Connective tissue encasing a myofiber

endoplasmic reticulum—Network of membranes within a cell

epididymis—Final sperm storage site

epimysium—Connective tissue encasing whole muscles

equestrian—Event involving both horse and rider

estimated breeding value (EBV)—Allows comparisons within a group, herd, or flock using heritability

estrogens—Female sex hormones

estrous cycle—Corresponds with sequential ovarian structures

ethology—Study of animal behavior

ewes—Female sheep

expected progeny difference (EPD)—Sophisticated selection tool that compares animals using data from related animals inside and outside the herd

F

farrier—Hoof caretaker

farrow to finish—Hogs owned from birth to slaughter

feeder calf—Young beef calf just weaned

feral—Wild animal that was once domesticated and has then returned to the wild

fine wool breed—Sheep breeds with excellent wool quality

finish—Fat cover

finisher—Person operating a feedlot

finishing cattle—Feedlot beef cattle

flat nipples—Functional, but not prominent nipples in swine

Flehman response—Curled upper lip in males after smelling the genitalia of the females

flight zone—Area surrounding an animal that, if invaded, causes the animal to move away

flock EPDs (FEPDs)—Sheep EPD calculated on animals within a flock

flushing—Increasing energy content of ewe's diet before breeding

follicle—Developing egg on an ovary

formula pricing—Pre-negotiated price

founder—Inflammation of the hoof

free-stall housing—Allow dairy cows to enter or leave a stall at-will

freemartin—Heifer born twin to a bull

futures price—Cash price expected in the future

G

gametes—Animal sex cells

general purpose index (GPI)—Assigns equal weights to maternal, growth, and carcass traits in swine

generation interval—Time required to replace a generation

genes—Small coded pieces of DNA

genetics—Study of heredity

genomics—Study of large portions of the genome

genotype—Actual genetic code

gilts—Young female swine

gizzard—Grinds coarse particles in poultry

glucose—Energy in the blood stream, blood sugar

glycogen—Glucose stored in muscle tissue

golgi bodies—Cellular structures that assemble and package cellular products

grade—Unregistered

grants—Reasons why an animal with a lower placing might be better

grower (beef)—To grow animals until they are more mature for finishing; also known as stocker operations

H

hand mating—Supervised mating

haploid—Number of pairs of chromosomes

heifers—Young female cattle

heritability—Portion of observed animal performance resulting from genetics

heterosis—The increase in a performance trait that exceeds the average of the parents; also called hybrid vigor

heterozygous—Having one copy of two different genes

homozygous—Having two copies of the same gene

hothouse lamb—Lambs raised indoors

hydrochloric acid—Stomach acid

hyperplasia—Muscle cell division before birth

hypertrophy—Muscle cell enlargement

I

inbreeding—Mating of closely related animals

incomplete dominance—Both genes of a pair are expressed

independent producer—Not aligned with a larger business entity

individual data—Performance of a single animal

infundibulum—Funnel surrounding the ovary

ingestive behaviors—Eating behaviors

Institutional Meat Purchasing Specifications (IMPS)—Standardized meat cuts

intact—Uncastrated

integrated production—Few producers own many pigs and control feed and/or slaughter facilities

intensive rotational grazing—A grazing system in which cattle are allowed access to a paddock of grass for as little as a day or two

intermuscular fat—Fat between muscle bundles

internal fat—Fat surrounding the internal organs, kidneys, pelvic cavity, and the heart

intramuscular fat—Fat inside muscle bundles

J

jockeys—People who buy and sell livestock

K

keep-cull—Simulation of replacement animal selection in a judging contest

ketosis—Disorder associated with a negative energy balance during early lactation

kidney, pelvic, and heart fat (KPH)—Internal fat in beef carcasses

knock-kneed—Knees too close together (front view)

L

lactating—Giving milk

lateral suspensory ligament—Separates fore and rear udder

lean-to-fat ratio—Comparison of lean to fat in an animal's carcass

learned behaviors—Behaviors acquired over time

leptospirosis—Disease that causes severe reproductive failure

Leydig cells—Testicular cells that produce testosterone

lifetime net merit (LNM)—A dairy cow's ability to be profitable in her lifetime

line breeding—Mating of distantly related animals

linear classification—Utilizes heritability of type traits in dairy

Linneaus—He developed a classification system for living organisms

lip prehenders—Grasp forages with the lips

litter weight at 21 days (LW21)—Weight of a litter of pigs

longeing—Exercising horses using a long rope

lordosis—Pulsing of ears during standing heat in swine

lysosomes—Digest and recycle used molecules within a cell

M

macrominerals—Minerals needed in grams per day

maintenance behaviors—Behaviors basic to life

marbled—Having intramuscular fat

mastitis—Mammary gland infection

maternal breeds—Breeds that excel in maternal traits

maternal line index (MLI)—Combines NBA and LW21 in swine

maternal milk (MMEPD)—Measures milk production in beef

maturity—Adulthood

median suspensory ligament—Divides udder in half from side to side

medium wool breed—Breeds with average fleece quality

meiosis—Sex cell division

metritis—Uterine infection

microminerals—Minerals needed in tiny amounts per day

milk fever—Results from calcium imbalance

mitochondria—Manufacture energy for the cell; also called the powerhouse of the cell

mitosis—Non-sex cell division

monogastric—Having one simple stomach

multiple trait indexes—Selects for several traits at one time

muscle fatigue—state of muscle exhaustion

mycoplasmal pneumonia—Disease of swine with few signs except coughing

myoblasts—Immature muscle cells that cannot divide

myofiber—Mature muscle cell

myosin—Thick muscle protein that helps in muscle contraction

myotube—Fused myoblasts

N

negative reinforcement—Punishment for a wrong behavior

net energy—Energy available for animal use

niche market—Specialized market

nucleus—Controls all cell activity

number born alive (NBA)—Number of live pigs born per litter

O

omasum—The third compartment of the ruminant stomach that absorbs water from the chyme; also known as manyplies

out breeding—Mating of animals that are not related; also known as crossbreeding

ovary—Primary female reproductive organ

oviduct—Connects ovary with uterus

P

parlor system—Cows come here to be milked

parvovirus—An infection in swine that results in failure of sows to come into heat, farrowing of small litters, and giving birth to mummified piglets

paternal breeds—Breeds excelling in growth and muscling

penis—Male copulatory organ

peptidases—Pancreatic enzymes that degrade proteins

percent difficult births in heifers (%DBH)—Indicates ease of calving in dairy

performance testing—Measuring growth rate and carcass composition data from a contemporary group

perimysium—Connective tissue encasing myofiber bundles

peristalsis—Muscle contractions that move food through the digestive system

phenotype—Portion of genetic code expressed by an animal

pigeon-toed—Front feet turned in (front view)

pin nipples—Small, undeveloped nipples in swine

placenta—Uterine membrane

polled—Naturally hornless

porcine stress syndrome—Condition causing pale, soft meat in pigs

positive reinforcement—Reward for a correct behavior

postmortem—After death

postnatal—Happening after birth

posty—Rear legs too straight (side view)

predators—Animals that kill other animals for food

predicted transmitting ability (PTA)—Dairy equivalent of an EPD

predicted transmitting ability for type (PTAT)—Overall ability to genetically transmit type traits in dairy

preliminary yield grade (PYG)—Unadjusted yield grade calculated from fat thickness at the 12th rib in beef

presumptive myoblasts—First muscle cells that can divide

price floor—Lowest possible price

primal cuts—Large carcass sections

production type index (PTI)—General purpose index for non-Holstein dairy animals, including type

progeny—Offspring

proventriculus—True stomach of poultry

PRRS (Porcine Reproductive and Respiratory Syndrome)—Viral disease with two components namely the reproductive component and the respiratory component

pseudorabies—Swine disease abbreviated PRV, characterized by death of large numbers of baby pigs

PTA$ for cheese yield (CY$) — Predicted dollar value of progeny for cheese yield in dairy

PTA$ for milk and fat (MF$) — Predicted dollar value of progeny for milk and fat in dairy

PTA$ for milk, fat, and protein (MFP$) — Predicted dollar value of progeny for milk and contents in dairy

Q

qualitative traits — Traits controlled by one gene pair that cannot be altered by environment

quality grades — USDA estimate of quality in beef and lamb

quantitative traits — Traits controlled by several or many gene pairs and subject to environmental alterations

R

rams — Intact male sheep

ratio — Compare animals using a percentage approach for a single trait within a contemporary group

reasons — Explanation for placing logic

replacement heifers — Young breeding beef or dairy females

rete testis — First ducts that collect immature sperm cells

reticulorumen — Rumen plus reticulum

ribosomes — Manufacture cellular proteins

rigor mortis — Muscle stiffness after death

rotaterminal crossbreeding system — Requires three maternal breeds and one terminal sire breed

ruminant — Animal with unique four-part stomach

S

sarcomere — Small unit of myofiber that contracts

sarcoplasmic reticulum — Calcium reservoir for muscle cells

scenario — Production situation

scur — Small unattached horn

seasonal breeder — Breed only at certain times of the year

seedstock — Breeding stock

selection intensity — Percentage of top animals retained for breeding

seminiferous tubules — Site of male meiosis

settle — Become pregnant

shipping fever (beef) — A group of respiratory diseases also known as bovine respiratory disease complex (BRDC)

sickle-hocked — Rear legs too curved (side view)

silage — Fermented forage

simple carbohydrates — Starches and sugars, that can be easily digested even by monogastric animals

social behaviors — Behaviors involving more than one animal

somatic cell count — Measurement of a milk sample indicating facility cleanliness and the presence of mastitis

somatotropin (growth hormone) — Protein hormone present in all animals

sow productivity index (SPI) — Gauges maternal performance in swine

splay-footed — Front feet turned out (front view)

split-sex feeding — Barrows and gilts fed separately

spool joint — White front leg bone of a sheep carcass greater than one year of age

stanchion — Tie stall facility for dairy cattle

steers — Male cattle with testicles removed

stillborn — Born dead

subcutaneous fat — Fat under the skin

subprimal cuts — Smaller carcass sections

sulky — Cart

swine erysipelas—Swine disease with diamond-shaped patches on the skin

symbiotic—Mutually beneficial relationship

synthetic breed—New breed derived from planned matings

T

tack—Harnessing, riding, and grooming equipment

taxonomy—Structured classification system for animals

terminal crossbreeding—The most effective way of maximizing heterosis; all resulting offspring are destined for slaughter

terminal sire—Sire in a crossbreeding system where all progeny are marketed

terminal sire indexes (TSI)—Emphasizes growth/carcass traits in swine

testis—Primary male reproductive organ

thermoregulatory behaviors—Behaviors to maintain body temperature

three-breed rotational cross—Uses breeds A, B, and C in rotation

titin—Prevents muscles from overstretching

tongue prehenders—Grasp forages with the tongue

transition phrase—Phrase inserted between reasons when shifting from one pair to the next

tropomyosin—Protein that regulates muscle contraction in conjunction with troponin

troponin—Protein that prevents muscle contraction when muscles are not stimulated

type classification—Dairy type score based on 100

type-production index (TPI)—General purpose index for Holsteins, including type

U

urethra—Tube connecting bladder to penis opening

USDA meat graders—Assign quality and yield grades

USDA meat inspector—Ensure meat safety and wholesomeness

uterus—Womb

V

vagina—Connects cervix with vulva

vas deferens—Tube from epididymis to urethra

villi—Finger-like projections of small intestine

vitelline block—Prevents second sperm from entering an egg

volatile fatty acids—Compounds absorbed by rumen lining

vulva—Visible portion of female reproductive tract

W

wean to finish (swine)—Twelve-pound weaned pigs moved to a finishing floor where they remain until marketing

weaning—Removal from the dam

weaning weight EPD (WW EPD)—Expected progeny difference for weaning weight in beef

wethers—Male sheep with testicles removed

winter—Maintain brood cows in cold weather

Y

yearling weight EPD (YW EPD)—Expected progeny difference for yearling weight in beef

yield grades—USDA estimate of cutability

Z

zona pellucida—Barrier surrounding egg

zygote—Diploid fertilized egg

BIBLIOGRAPHY

Baker, M., and R. E. Mikesell. 2005. *Animal Science Biology and Technology.* Pearson Prentice Hall Interstate, Upper Saddle River, NJ.

Boggs, D. L., R. A. Merkel, M. E. Doumnit, and K. W. Bruns. 2006. *Live Animal Carcass Evaluation and Selection Manual.* 6th ed. Kendall/Hunt Publishing Co., Dubuque, IA.

Bryan, K. A. 1994. *Pennsylvania State 4-H Livestock Judging Manual.* College of Agricultural Sciences, Cooperative Extension, The Pennsylvania State University.

Bush, L. J. 1979. *Livestock Feeding.* Oklahoma State University, Stillwater.

Campbell, J. R., and J. F. Lasley. 1999. *The Science of Animals that Serve Humanity.* 3rd ed. McGraw-Hill Book Co., New York.

Campbell, J. R., and R. T. Marshall. 1978. *The Science of Providing Milk for Man.* McGraw-Hill Book Co., New York.

Cheeke, P. R. 1993. *Impacts of Livestock Production: On Society, Diet/Health and the Environment.* Interstate Publishers, Inc., Danville, IL.

Cole, H. H., and M. Ronning, eds. 1974. *Animal Agriculture: The Biology of Domestic Animals and Their Use by Man.* W. H. Freeman and Co., San Francisco.

Cullison, A. E., and R. S. Lowry. 1987. *Feeds and Feeding.* 4th ed. Pearson Prentice Hall, Upper Saddle River, NJ.

Cunha, T. J. 1977. *Swine Feeding and Nutrition.* Academic Press, New York.

Ensminger, M. E. 1987. *Beef Cattle Science.* 6th ed. Interstate Publishers, Inc., Danville, IL.

Ensminger, M. E. 1990. *Horses and Horsemanship.* 6th ed. Interstate Publishers, Inc., Danville, IL.

Ensminger, M. E. 1991. *Animal Science.* 9th ed. Interstate Publishers, Inc., Danville, IL.

Ensminger, M. E. 1992a. *Animal Science Digest.* Interstate Publishers, Inc., Danville, IL.

Ensminger, M. E. 1992b. *Stockman's Handbook Digest.* Interstate Publishers, Inc., Danville, IL.

Ensminger, M. E., and R. O. Parker. 1984. *Swine Science.* 5th ed. Interstate Publishers, Inc., Danville, IL.

Ensminger, M. E., and R. O. Parker. 1986. *Sheep & Goat Science.* 5th ed. Interstate Publishers, Inc., Danville, IL.

Etgen, W. M., and P. M. Reaves. 1987. *Dairy Cattle Feeding and Management.* 7th ed. John Wiley & Sons, New York.

Etherton, T. 1990. Animal Growth and Development. Animal Science 514. The Pennsylvania State University. Class notes.

Gerrard, D. E., and A. L. Grant. 2006. *Principles of Animal Growth and Development.* Kendall/Hunt Publishing Co., Dubuque, IA.

Gillespie, J. R. 2010. *Modern Livestock and Poultry Production.* 7th ed. Delmar Cengage Learning, Clifton Park, NY.

Greer, J. G., and J. K. Baker. 1992. *Animal Health: A Layperson's Guide to Disease Control.* 2nd ed. Interstate Publishers, Inc., Danville, IL.

Herren, R. V. 2007. *The Science of Animal Agriculture.* 3rd ed. Delmar Cengage Learning, Clifton Park, NY.

Holstein-Fresian Association of America. 1984. "When You Talk About Holsteins. . . Use the Right Words." Pamphlet, Battleboro, VT.

Krider, J. L., J. H. Conrad, and W. E. Carroll. 1982. *Swine Production.* 5th ed. McGraw-Hill Book Co., New York.

Lasley, J. F. 1978. *Genetics of Livestock Improvement.* Prentice-Hall Inc., Englewood Cliffs, NJ.

Lawhead, J. B., and M. Baker. 2005. *Introduction to Veterinary Science.* Delmar Cengage Learning, Clifton Park, NY.

Lee, J. S., C. Embry, J. Hutter, J. Pollock, R. Rudd, L. Westrom, and A. Bull. 1996. *Introduction to Livestock and Poultry Production.* Interstate Publishers, Inc., Danville, IL.

Lister, D., D. N. Rhodes, V. R. Fowler, and M. F. Fuller, eds. 1974. *Meat Animals: Growth and Productivity.* Plenum Press, New York.

McLaren, G. A. 1976. *Biochemistry of Animal Nutrition.* Division of Animal and Veterinary Sciences, West Virginia University, Morgantown.

Moody, E. G. 1991. *Raising Small Animals.* Farming Press Books, Ipswich, UK.

National Agricultural Statistics Service. *2007 Census of Agriculture.* Available at http://www.agcensus.usda.gov/.

Neuman, A. L. 1977. *Beef Cattle.* 7th ed. John Wiley & Sons, New York.

Pond, W. G., and J. H. Maner. 1984. *Swine Production and Nutrition.* AVI Publishing Company, Inc., Westport, CT.

Purebred Dairy Cattle Association. 1994. *Dairy Cow Unified Score Card.*

Romans, J. R., W. J. Costello, C. W. Carlson, M. L. Greaser, and K. W. Jones. 1994. *The Meat We Eat.* 14th ed. Interstate Publishers, Inc., Danville, IL.

Schillo, K. K. 2009. *Reproductive Physiology of Mammals: From Farm to Field and Beyond.* Delmar Cengage Learning, Clifton Park, NY.

Schneider, B. H., and W. P. Flatt. 1975. *The Evaluation of Feeds Through Digestibility Experiments.* The University of Georgia Press, Athens.

Taylor, R. E. 1984. *Beef Production and the Beef Industry: A Beef Producers Prospective.* Macmillan Publishing Co., New York.

Taylor, R. E., and R. Bogart. 1988. *Scientific Farm Animal Production: An Introduction to Animal Science.* Macmillan Publishing Co., New York.

Williams, W. F., and T. T. Stout. 1964. *Economics of the Livestock-Meat Industry.* Macmillan Publishing Co., New York.

INDEX

2N number. *See* Diploid number
8-cell embryo, 74

A

A band, 42
Abomasum, 62
Actin, 42, 43
Adipocytes, 29
Adipose tissue, 29, 209
Aerobic exercise, 44
Age determination, of horses,
    279–281
Agro-terrorism, 289
Alfalfa hay, 277
All-in all-out production, 138
Alleles, 89
American Angus Association,
    111
American Horse Council (AHC),
    260, 290
American Paint Horse
    Association, 265
American Quarter Horse
    Association (AQHA),
    262
American Saddlebred Horse
    Association, 270
American Saddlebreds, 270
American Sheep Industry
    Association (ASI), 254
Amino acids, 53, 54
Anaerobic exercise, 44
Anaphase, 8, 10

Anaplasmosis, 183–184
Androgens, 27, 33
Anestrus, 70
Angus Heifers scenarios,
    326–327
Animal behaviors, 104
    eliminative, 105
    individual maintenance
        behaviors, 105–106
    ingestive, 105
    learned behaviors, 109–110
    and livestock movement,
        112–114
    sleeping, 105
    social behaviors, 106–109
    thermoregulatory, 105
Animal rightists, 114–115
Animal rights, 114–115
Animal taxonomy, 2, 12
    cattle, 15
    horses, 16–17
    sheep, 16
    swine, 14–15
Animal welfare, 114–115
Animal welfarists, 114–115
Animalia, 13
Anthelmintic, 283
Apoptosis, 8
Appaloosa, 267
Appaloosa Horse Club, 267
Arabian breed, 264
Arabian Horse Registry, 264
Artificial insemination (AI), 98,
    205, 276

Artiodactyla, 13
Association of Racing
    Commissioners
    International, Inc., 291
Atrophic rhinitis, 143
Average daily gain (ADG), 314,
    320
Ayrshires, 202

B

Baby pig processing, 135
Backcrossing, 130
Backfat at 250 pounds (BF), 320
Backgrounding, 179
Bakewell, Robert, 88, 89
Baldies, 166
Band of broodmares, 275
Barley, 277
Barrows, 24
Beef
    breeding evaluation, 389–394
    judging, 338
        beef terminology and
            reasons, 346–353
        judging beef carcasses, 346
        quality grading, 341–346
        yield grading, 339–341
    market evaluation, 369–376
Beef cattle management, 160
    beef production cycle,
        176–180
    breeding systems, 176
    breeds, 164

Beef cattle management
(*continued*)
    Black Angus, 166, 167
    Brahman, 168–169, 170
    Charolais, 170–171
    Chianina, 171–172
    Gelbvieh, 172, 173
    Herefords, 165–166, 167
    Limousin, 172–173
    Longhorns, 164–165
    Maine-Anjou, 173–174
    Red Angus, 167–168
    Salers, 174–175
    Shorthorns, 165
    Simmental cattle, 175
  foot rot, 187
  gastrointestinal diseases, 187
  housing requirements,
    187–188
  nutrition, 180–182
  parasites
    anaplasmosis, 183–184
    coccidiosis, 184
    flies, 183
    mange, 184–185
    ringworm, 184
    stomach worms, 185
  pinkeye, 187
  reproductive diseases, 186
  respiratory diseases, 185
  scenarios, 324–327
  tuberculosis (TB), 187
Beef chucks, 348–350
Beef industry, 160
  career opportunities, 192–193
  issues, 190–192
  marketing, 188–189
  organization, 190
  supporting organizations, 193
Beef production cycle, 176–180
Behavior and livestock
    movement, 112–114
Behavioral genetics, 111–112
Belgian, 272
Belgian Draft Horse
    Corporation of America,
    272
Berkshire, 122–123
Beta agonists, 34
Beta carotene, 203

Binomial nomenclature, 13–14
Binomial system. *See* Binomial
    nomenclature
Biopryn® test, 75
Biosecurity, 145
Bioterrorism, 153
Birth weight EPD (BW EPD),
    321
Black Angus, 166, 167
Black baldies, 166
Blind nipples, 394
Blood system, 251
Bloody body. *See* Corpus
    hemorrhagicum (CH)
Bluegrass, 277
Boars
  breeding evaluation, 397–398
Body condition, rebuilding, 209
Bos indicus, 15
Bos taurus, 15
Bovine respiratory disease
    complex (BRDC), 185
Bovine rhinotracheitis (IBR), 185
Bovine somatotropin (bST), 33
Bovine spongiform
    encephalopathy
    (BSE), 191
Bovine viral diarrhea (BVD), 185
Bowlegged, 372, 373
Box stall, 286–288
Brahman, 168–169, 170
Braising, 338
Break joint, 361
Breeding
  of dairy cattle, 205–206
  of horses, 274–276
    artificial insemination (AI),
      276
    biology of, 275
    gestation period for, 276
    hand mating, 275
    large-scale, 274–275
Breeding animal evaluation, 389
  beef, 389
    bulls, 393–394
    heifers, 390–393
  performance data, 402
  sheep, 398
    ewes, 398–401
    rams, 401–402

  swine, 394
    boars, 397–398
    gilts, 394–397
Breeding systems, 130
  backcrossing, 94–96, 130
  in beef cattle, 176
  crossbreeding, 95–96
  inbreeding, 94–95
  livestock selection, 332
  rotaterminal crossbreeding,
    131–132
  in sheep, 238–240
  terminal crossbreeding,
    132–133
  three-breed rotational
    crossbreeding, 131
Breeds
  of beef cattle
    Black Angus, 166, 167
    Brahman, 168–169, 170
    Charolais, 170–171
    Chianina, 171–172
    Gelbvieh, 172, 173
    Herefords, 165–166, 167
    Limousin, 172–173
    Longhorns, 164–165
    Maine-Anjou, 173–174
    Red Angus, 167–168
    Salers, 174–175
    Shorthorns, 165
    Simmental cattle, 175
  of sheep, 232
    Columbia, 235, 236
    Corriedale, 237
    Dorset, 233–234
    Hampshire, 234
    Polypay, 235, 236
    Rambouillet, 234–235
    Shropshire, 238
    Southdown, 237
    Suffolk, 232–233
    Texel, 238
  of swine, 121
    Berkshire, 122–123
    Chester White, 123
    Chinese strains, of swine,
      129
    Duroc, 123
    Hampshire, 124, 125
    Herefords, 128

Jersey Reds, 123–124
Landrace, 125
Pietrains, 128, 129
Poland Chinas, 126
Spotted Swine, 126, 127
Tamworth, 128
Bromegrass, 277
Brood cows, 190
Brown Swiss, 201, 202
Brucellosis, 186
Buck-kneed, 372, 373
Bucks. *See* Rams
Bulls
breeding evaluation, 393–394
Buttons, 342
Bypass fat, 212

C

Calcification, 31
Calcium, 215
Calf hutches, 216
Calf-kneed, 372, 373
Calves, diets for, 207, 208
Calving, 79
Canola meal, 277
Capacitation, 72–73
Carbohydrates, 56
Cardiac muscle, 27
Career opportunities
in beef industry, 192–193
in dairy industry, 224
in horse industry, 289
in sheep production, 254
in swine industry, 153–155
Carnivora, 13
Cartilage, 31
Cash price, 150
Castrated, 27
Cattle grubs, 183
Cattle, taxonomy of, 15
Cecum, 59, 277
Cell cycle, 6
Cell membrane, 5
Cell theory, 5
Cellular biology, 2
structure and function, 5
cells, 10
cellular reproduction, 6–10
organs, 11

systems, 11–12
tissues, 10
Cellular reproduction, 6
meiosis, 8–10
mitosis, 8
Centromere, 7
Cervix, 71
Charolais, 170–171
Check-off funds, 253
Check-off program, 155
Chester White, 123
Chianina, 171–172
Chinese strains, of swine, 129
Choice/select spread, of
carcasses, 348
Chromosomes, 5–6
Chyme, 58
Class, 13
Clostridia, 247–248
Clostridium, 186
Clostridium Chauvoei, 247
Clostridium Perfringens, 247
Clostridium tetani, 247, 286
Clover, 277
Clydesdales, 273
Coarser wools, 251
Coccidiosis, 184, 247
Codominance, 91
Coggins Test, 284
Colic, 283
Collagen, 31, 42
Colored breeds, 409
Colostrum, 75, 207
Columbia, 235–236
Commercial cattle, 164
Compensatory gains, 188
Complete dominance, 91
Complex carbohydrates, 56
Computer controlled milking
inflations, 218
Computerized feeding system,
211
Conception, 23
Conditioning, 110
Contemporary groups, 93, 94
Contestants
cooking different cuts, 338
in meat judging contests, 338
Contestants score, 307–308
Continental breed, 170

Cooking different cuts,
of meat, 338
Cooperatives, 219
Corn, 211, 277
Corn Belt, 121
Corpus hemorrhagicum (CH), 69
Corpus luteum (CL), 69–70
Corriedales, 237
Cottonseed meal, 277
Count system, 251
Country of origin labels
(COOL), 191
Cow-calf operations, 190
Cow diets, 209–213
Cowhocked, 372, 373
Coyotes, 253
Creep feeding, 243
Crimped corn, 244
Crisscrossing. *See* Backcrossing
Crop, the, 60
Cross-fostered, 137
Crossbreeding, 95–96
Crutching, 246
Cull, animals, 302
Cutability, 333
Cuts of meat, 307–308,
333–338
cooking different, 338
from hindquarters, 336
from loin, 336
from lower limbs, 336
from neck, 336
primal cuts, 333
retail cuts, 338
from shoulder, 336
subprimal cuts, 337
Cytoplasm, 6

D

Dairy breeds, 199
Ayrshire, 202
Brown Swiss, 201, 202
Guernsey, 203
Holstein, 200
Jersey, 201
Milking Shorthorn, 203–204
red and white, 203–204
Dairy bull calves, 205
Dairy calf nutrition, 207–213

Dairy cattle judging
  dairy cows, judging, 414
    dairy strength, 417–419
    frame, 415–417
    rear feet and legs, 419
    sample reasons, 423–424
    udder, 420–423
  dairy heifers, judging, 424
    sample reasons, 424–425
  linear classification, 413–414
  type classification, 409–413
Dairy cattle management, 196
  breeding, 205–206
  breeds, 199
    Ayrshire, 202
    Brown Swiss, 201, 202
    Guernsey, 203
    Holstein, 200
    Jersey, 201
    Milking Shorthorn, 203, 204
    red and white, 203–204
  housing, 216–218
  milking process, 206
  nutrition, 207–213
  parasites and diseases, 213
    displaced abomasum
      (DA), 215
    ketosis, 214
    mastitis, 213–214
    metritis, 216
    milk fever, 215
    retained placenta, 216
  records of milk production,
    218–219
Dairy check-off program, 224
Dairy Cow Unified Score Card,
    409, 410–411, 412,
    413, 415
Dairy cows
  feeding lactating in, 209
  judging, 414
    dairy strength, 417–419
    frame, 415–417
    rear feet and legs, 419
    sample reasons, 423–424
    udder, 420–423
Dairy farms, 220–221
Dairy food scientists, 224
Dairy heifers, judging, 424
  sample reasons, 424–425

Dairy Herd Improvement
    Associations (DHIA), 219
Dairy housing, 216–218
  calf hutches, 216
  free-stall housing, 217
  stanchion barns, 216
  tie-stall, 216
Dairy industry, 196
  career opportunities, 224
  industry organization,
    220–221
  issues, 221–223
  milk marketing, 219–220
  overview, 199
  supporting organizations,
    224–225
Dam breeds, 121
Dark, firm, dry (DFD), 46, 47
Days to 250 pounds (DAYS), 320
Decision-making process, 299
Developer diets, 139, 140
Diestrus, 70
Digestible energy (DE), 57
Digestion, biology of, 50
  digestion process, 58–59
  digestive systems, 60
    cattle, 61–63
    goats, 61–63
    horses, 63–64
    poultry, 60–61
    sheep, 61–63
    swine, 60
  energy metabolism, 57–58
  essential nutrients, 53
    carbohydrates, 56
    fats, 56–57
    minerals, 55
    protein, 53–54
    vitamins, 54–55
    water, 53
Diploid number, 9
Diseases
  in dairy cattle
    displaced abomasum
      (DA), 215
    ketosis, 214
    mastitis, 213–214
    metritis, 216
    milk fever, 215
    retained placenta, 216

  in horses, 283
    distemper, 285–286
    encephalomyelitis, 283
    equine infectious anemia
      (EIA), 284
    equine influenza, 284
    founder, 284–285
    potomac horse fever, 285
    rabies, 285
    tetanus, 286
    West Nile Virus, 286
  in sheep, 247
    foot rot, 248
    pneumonia, 248
    scrapie, 253
    sore mouth, 248–249
  in swine, 142
    atrophic rhinitis, 143
    leptospirosis, 143–144
    mycoplasmal pneumonia, 144
    parvovirus, 144
    PRRS (porcine reproductive
      and respiratory
      syndrome), 145
    pseudorabies (PRV), 145
    swine erysipelas, 143
Displaced abomasums (DA), 215
Distemper, 285–286
DNA (Deoxyribonucleic Acid), 5
Dorset, 233–234
Down breeds, 232
Draft horse breeds, 272–273
  Belgian, 272
  Clydesdales, 273
  Percheron, 272–273
Drenching, 247
Dressage, 290
Dressing percentage, 189
Dry cows, 210
Dry matter basis, 209–210
Duroc, 123
Dust generation, 223
Dystocia, 176

E

E. coli, 186
Eared cattle, 169
Economically important
    tissues, 25

bone development, 31–32
fat development, 29–31
muscle development, 27–29
Efferent ducts, 72
Eliminative, 105
Embryo, 73
Encephalomyelitis, 283
Endomysium, 41
Endoplasmic reticulum, 6
Energy metabolism, 57–58
Epididymis, 72
Epimysium, 42
Epiphyseal growth plate, 31
Equestrian events, 275
Equine infectious anemia
(EIA), 284
Equine influenza, 284
Equus caballus, 16
Esophageal groove, in calves,
207
Essential nutrients, 53
carbohydrates, 56
fats, 56–57
minerals, 55
protein, 53–54
vitamins, 54–55
water, 53
Estimated breeding value
(EBV), 317
Estrogen, 33, 69, 71
Estrous cycle, 70
Estrus, 70, 82
Ethology, 102
animal behaviors, 104
individual maintenance
behaviors, 105–106
learned behaviors,
109–110
social behaviors, 106–109
animal rights, 114–115
animal welfare, 114–115
behavior and livestock
movement, 112–114
behavioral genetics,
111–112
Ewes, 28, 80
breeding evaluation,
398–401
Expected progeny differences
(EPD), 317

Family, 13
Farrier (hoof caretaker), 260
Farrow to finish, 151
Fast, white fibers, 43–43, 45
Fat cover, 371
Fats, 56–57
Federal Milk Marketing
Orders, 219
Feed additives, 212
Feeder calf, 176, 205
Feeding
dry matter basis, 209–210
feed additives, 212
forages in, 211
high feed intake, reasons for,
210
in lactation cow, 211
nutrients, 210
nutritional needs of, 209–210
Feedlot cattle, 182
Female reproductive system,
69–71
Fencing for horses, 288
Feral, 14
Fertilization
physiological changes in,
72–73, 74
Fescue, 277
Fetus, 73, 74
Fine-wool breed, 234. *See also*
Rambouillet
Fine wools, 251
Finish, 371
Finisher, 190
Finisher diet, 139, 140
Finishing cattle, 181, 182, 188
Flat nipples, 394–395
Flehmen response, 108
Flies, 183
Flight zone, 112
Flock EPDs/FEPDs, 322
Flushing, 241
Foal heat, 82
Foals breeding, 277
Follicle, 69
Foot and hoof health of horses,
281
Foot rot, 187, 248

Forage quality, 211
Forages, 62
Formula price, 151
Founder, 284–285
Franklin, Rosalind, 6
Free-stall housing, 217
Freemartin, 80
Freshening, 79
Fungi, 13
Futures price, 150–151

G1 period, 6, 7
G2 period, 6, 7
Gametes, 8
Gastrointestinal diseases, of beef
cattle, 187
Gelbvieh, 172, 173
Gender determination, 89–90
General purpose index (GPI),
320
Generation interval, 99
Genes, 6, 89
Geneticists, 318
Genetics, 86
breeding systems, 94–96
gender determination, 89–90
genomics, 99
genotype, 90
heritability, 96–97
Mendelian genetics, 89
phenotype, 90
qualitative traits, 90–92
quantitative traits, 92–94
selection intensity, 97–99
Genomics, 99, 319–320
Genotype, 90
Genus, 13
Gestation diet, 139, 140–141
Gestation period
for horses, 276
of sheep, 241
Gilts, 28
breeding evaluation, 394–397
Gizzard, 60
Glucose, 32, 43
Glycogen, 43
Golgi bodies, 6
Grade animals, 205

Grain mix ingredients, 211
Grants, 303
Grid pricing systems, 191
Gross energy (GE), 57
Grower diets, 139, 140
Grower operations, 190
Growth and development,
          biology of, 20
      economically important
            tissues, 25
          bone development, 31–32
          fat development, 29–31
          muscle development,
                27–29
      growth curve, 23–25
      muscle to fat ratio, 32
          beta agonists, 34
          genetic selection, 34–35
          sex hormones, 34
          somatotropin (ST), 32–33
Growth curve, 23–25
Guernsey, 203

H

H zone, 43
Hampshire, 124, 125, 234
Hand mating, 275
Haploid number, 8
Heavy horse breeds. *See* Draft
          horse breeds
Heel flies, 183
Heifer, 27, 77
      breeding evaluation, 390–393
      diets, 207
      tissue deposition in, adipose 209
      weight of large breed, 208
      weight of smaller breed, 209
Herefords, 128, 165–166, 167
Heritability, 96–97
Heterosis, 95, 96
Heterozygous, 92
Holstein, 200, 204
Holstein Association, 409
Holstein-Friesians. *See* Holstein
Homozygous, 92
Hoof trimming, 246
Hormones, 12
Horse age determination,
          279–281

Horse breeds, 261–274
      classification of, 261–262
      draft, 272–273
      light, 262–271
      Miniature, 273–274
      Ponies, 273–274
Horse diet, 279
Horse industry
      career opportunities in, 289
      industry organization, 288
      issues, 288–289
      marketing, 288
      overview, 260–261
      supporting organizations,
            290–291
Horse industry organization,
          288
Horse management, 258,
          279–282
      breeding, 274–276
      breeds, 261
          draft horse breeds,
                272–273
          light horse breeds, 262–271
          ponies and miniature
                horses, 273–274
      diseases, 283
          distemper, 285–286
          encephalomyelitis,
                283–284
          equine infectious anemia
                (EIA), 284
          equine influenza, 284
          founder, 284–285
          potomac horse fever, 285
          rabies, 285
          tetanus, 286
          West Nile Virus, 286
      housing, 286–288
      nutrition, 276–279
      parasites, 282
          flies, 282
          mange, 282
          ringworm, 282
          ticks, 283
          worms, 283
      taxonomy of, 16–17
Horse marketing, 288
Horse racing, 291
Horse stables, 289

Hothouse lambs, 240. *See also*
          "Out of season" lambs
Houseguest bacteria, 62
Housing
      horse, 286–288
      sheep, 249–250
      swine, 145–148
Housing requirements, for beef
          cattle, 187–188
Hybrid vigor. *See* Heterosis
Hydrochloric acid, 58
Hyperplasia, 41
Hypertrophy, 41

I

I band, 42
Ideal slaughter point, 25, 26
Identification methods of horses,
          279–281
Inbreeding, 94–95
Incomplete dominance, 91
Independent producers, 120
Individual data for animal
          industries, 320–323
      beef, 321–322
      dairy, 322–323
      sheep, 322
      swine, 320–321
Individual performance data,
          314–315
Infundibulum, 70
Ingestive behaviors, 105
Institutional Meat Purchasing
          Specifications
          (IMPS), 337
Intact, 27
Integrated production system, 120
Intensive pasture management,
          221
Intensive rotational grazing, 180
Intermuscular fat, 29
Internal fat, 29
International Dairy Foods
          Association (IDFA), 225
Interphase, 6–7
Intramuscular fat, 29, 31
Issues
      in beef industry, 190–192
      in dairy industry, 221–223

animal welfare concerns, 223

environmental issues, 223

inheritance tax, 222

loss of infrastructure, 222

vacillated milk prices, 222

in horse industry, 288–289

shortage of horse veterinarians, 289

speculation, 289

unwanted horse coalition, 289

in sheep industry, 252–254

importation of lamb, 252

predation, 253

scrapie tag system, 253–254

in swine industry, 151–153

J

Jersey, 201

Jersey Reds, 123–124

Jockeys, 254

Johne's disease, 186

Judging contests, 296, 298–299

format of, 299–302

livestock evaluation contests, 308

oral/written reasons, 302

quiz bowl contests, 308

reason components, set of, 302–307

accuracy, 302, 303

correct terminology, 304

delivery, 304–307

organization, 303–304

scoring classes in, 307–308

skillathon contests, 308

system of cuts in, 307–308

K

Keep–cull class, 302

Ketosis, 214

Kidney, pelvic, and heart fat (KPH), 340

Kingdoms, 13

Knock-kneed, 372, 373

L

Lactating cows, 177

Lactation

diet, 139, 141

physiological changes in, 75

Lamb

judging, 359

carcasses, 360–361

lamb terminology and reasons, 361–363

primal cuts, 362

market evaluation, 383–389

Lamb Promotion, Research and Information Board, 253

Laminitis. *See* Founder

Landrace, 125

Lasalocid, 182, 207

Lateral suspensory ligament, 420

Lean-to-fat ratio, 333

Learned behaviors, 104, 109–110

Leg faults, of swine, 380

Leptospirosis, 143–144, 186

Leydig cells, 71

Lice, 142

Lifetime net merit (LNM), 319

Light horse breeds, 262–271

American Saddlebreds, 270

Appaloosa, 267

Arabian breed, 264

Morgan, 269

Paint Horses, 264–265

Palomino, 270–271

Pinto, 269

Quarter Horse, 262–263

Standardbreds, 266

Tennessee Walking Horse, 267–268

Thoroughbreds, 263

Limousin, 172–173

Limousin Bulls scenarios, 325–326

Line breeding, 94

Linear classification, 413–414

Linnaeus, Carolus, 12, 13

Linseed meal, 277

Lip prehenders, 105

Litter weight at 21 days (LW21), 320

Livestock evaluation contests, 308

Livestock judging, 366

breeding animal evaluation, 389

beef, 389–394

performance data, 402

sheep, 398–402

swine, 394–398

market animal evaluation, 368

beef, 369–376

lambs, 383–389

swine, 376–383

Longeing, 281

Longhorns, 164–165

Lordosis, 76

Lysosomes, 6

M

Macrominerals, 55

Mad cow disease, 191

Maine-Anjou, 173–174

Maintenance behaviors. *See* Individual maintenance behaviors

Male reproductive organ, 71–72

Mammalia, 13

Mange, 142, 184–185

Manyplies, 62

Marbling, 30, 31, 319, 341, 343–345

category of, 344

score of, 345

Mares breeding, 277

Market animal evaluation, 368

beef, 369–376

lambs, 383–389

swine, 376–383

Market lambs, 239

Market price of wool, 251

Marketing

of cattle, 188–189

cash price, 150

formula price, 151

futures price, 150–151

of hogs, 148

of horses, 288

of lambs, 250–251

Marketing (*continued*)
  of milk, 219–220
  of sheep, 250–251
  of swine, 148–151
Mastitis, 213–214
Maternal breeds, 121, 239
Maternal indexes, 316
Maternal line index (MLI), 320
Maternal milk (MM EPD), 321
Maturity, 23
Meat biology, 38, 40, 45
  quality
    differences in, 46–47
    improvement, 47
Meat evaluation, 333
Meat quality defined, 332–333, 338
Meats judging, 330, 333
  beef judging, 338
    beef terminology and reasons, 346–353
    judging beef carcasses, 346
    quality grading, 341–346
    yield grading, 339–341
  contest, 337
  cooking different cuts, 338
  cuts of meat, 333–338
  lamb judging, 359
    lamb carcasses, 360–361
    lamb terminology and reasons, 361–363
  meat quality defined, 332–333
  pork judging, 354
    pork carcasses, 355–359
  sample beef carcass reasons, 353–354
Mechanical washing, 218
Median suspensory ligament, 420
Medium-wool breeds, 232. *See also* Suffolks; Down breeds
Meet breeds, 239
Meiosis, 8–10
Meiosis I, 9, 10
Meiosis II, 9, 10
Mendel, Gregor, 88, 89
Mendelian genetics, 89
Metabolizable energy (ME), 57
Metaphase, 8, 10

Metestrus, 70
Metritis, 216
Microminerals, 55
Micron system, 251
Milk, fat, and protein combined (MFP$), 323
Milk fever, 215
Milk let-down, 110
Milk marketing, 219–220
Milk price, 219
Milk price floor, 221
Milk production
  major factor for, 209
  peak lactation, 209
  percentage of, 199
  rebreeding, 209
  records of, 218–219
  stages of lactation, 209
Milking process, 206
Milking Shorthorn, 203, 204
Milo, 277
Minerals, 55, 279
Miniature horses, 273–274
Mitochondria, 6
Mitosis, 7, 8, 9
Molasses, 244
Monensin, 182, 207
Monera, 13
Monogastric, 60
Morgan, 269
Mucous discharge, 206
Multiple trait indexes, 316
Muscle biology, 38
  cells, 41–43
  hyperplasia, 41
  hypertrophy, 41
  myofibers, types of, 43–45
Muscle content, 332
Muscle fatigue, 43
Muscle to fat ratio, 32
  beta agonists, 34
  genetic selection, 34–35
  sex hormones, 34
  somatotropin (ST), 32–33
Muscling, 338
Mycoplasmal pneumonia, 144
Myoblasts, 27, 28
Myofiber, 28, 41–42
  types of, 43–45
Myosin, 42, 43

Myosin feet, 43
Myotubes, 28

N

N number. *See* Haploid number
National Animal Identification System (NAIS), 253, 289
National Cattlemen's Beef Association (NCBA), 193
National Dairy Board, 224
National Lamb Feeders Association, 255
National Milk Producers Federation (NMPF), 225
National Pork Board, 153, 155, 156
National Pork Producers Council (NPPC), 156
National sheep improvement center, 255
National Western Livestock Show, 308
Negative reinforcement, 109
Net energy (NE), 57
Niche market, 252
Non-lactating cows, 210
Note-taking format, 305
Nucleus, 5
Number born alive (NBA), 320
Nutrient Requirements of Dairy Cattle, 213
Nutrition
  beef cattle, 180–182
  dairy calf nutrition, 207–213
  nutrition, 276–279
    feeding horses, 278–279
    grains, 277
    hay, 276
    legumes, 277
    oats, 276
    pasture species, 277
    protein supplements, 277
    unfermented forages, 277
  sheep, 240–244
  swine, 138–141

O

Oats, 211, 276
Omasum, 62
Open shed, 286–288
Orchardgrass, 277
Order, 13
Organizations supporting horse industry, 290–291
Out breeding. *See* Crossbreeding
"Out of season" lambs, 240
Ovary, 69
Oviduct, 70
Ovis aries, 16
Oxytocin, 71–72, 74, 75

P

Paint Horses, 264–265
Pale, soft, exudative (PSE), 46, 47
Palomino, 270–271
Palpation, 79
Parainfluenza 3 (PI3), 185
Parasite infestations, 246
Parasites
    of beef cattle
        anaplasmosis, 183–184
        coccidiosis, 184
        flies, 183
        mange, 184–185
        ringworm, 184
        stomach worms, 185
    in horses, 282–283
        flies, 282
        mange, 282
        ringworm, 282
        ticks, 283
        worms, 283
Parlor system, 218
Parturition, 207
    physiological changes in, 74–75
Parvovirus, 144
Paternal breeds, 121
Peak of lactation, 75, 178, 209
Penis, 72
Peptidases, 58–59

Percent difficult births in heifers (%DBH), 322
Percheron, 272–273
Performance data, 312, 314
    for breeding evaluation, 402
    in genomics, 319–320
    for individual animal industries, 320–323
        beef, 321–322
        dairy, 322–323
        sheep, 322
        swine, 320–321
    in production situation, 324–327
    types of, 314–319
        estimated breeding value, 317–319
        expected progeny differences, 317–319
        indexes, 316
        individual data, 314–315
        predicted transmitting ability, 317–319
        ratios, 315–316
Performance testing, 166
Perimysium, 42
Perissodactyla, 13
Peristalsis, 58
Phenotype, 90
Phenotypic traits, 324
Phosphorus, 215
Phylum, 13
Physiological maturity, 342
Pietrains, 128, 129
Pigeon-toed, 372, 373
Piglet and nursery management, 135–138
Pin nipples, 395
Pinkeye, 187
Pinto, 269
Placing cards, 300–301
Plantae, 13
Pneumonia, 248
Poland Chinas, 126
Polands. *See* Poland Chinas
Polled Herefords, 166, 167
Polypay, 235
Ponies, 273–274
Porcine stress syndrome (PSS), 47

Pork carcasses judging, 354–359
    primal cuts in, 358
    sample, 359
    terminology for, 356
Pork Quality Assurance (PQA), 152–153
Positive reinforcement, 109
Postgastric fermenters, 277
Postmortem, 45
Postnatal, 41
Posty, 372
Potomac horse fever, 285
Predators, 253
Predicted transmitting ability (PTA), 318
Predicted transmitting ability for type (PTAT), 323, 414
Pregnancy
    physiological changes in, 74
Preliminary yield grade (PYG), 339
Presumptive, 27
Production type index (PTI), 322
Proestrus, 70
Professional Rodeo Cowboys' Association, 290
Progeny, 171
Progesterone, 70, 71
Programmed cell death. *See* Apoptosis
Prophase, 8
Prostaglandin, 70, 71
Protein, 53–54
Protista, 13
Proventriculus, 60
PRRS (porcine reproductive and respiratory syndrome), 145
Pseudorabies (PRV), 145
PTA$ for cheese yield (CY$), 323
PTA$ for milk and fat (MF$), 323
Punnett squares, 91
Purebred Brahmans, 169
Purebred Dairy Cattle Association, 409

Q

Qualitative traits, 90–92
Quantitative traits, 92–94
Quarter Horse, 262–263
Quiz bowl contests, 308

R

Rabies, 285
Rambouillet, 234–235
Rams
    breeding evaluation, 401–402
Ratio of performance data,
    315–316
Rebreeding, 209
Recombinant somatotropin
    (rST), 33
Red and white breed, 203–204
Red Angus, 167–168
Red mutation, 204
Reduction division, 10
Reed canarygrass, 277
Relaxin, 71, 74
Replacement ewes, 239
Replacement heifers, 176
Reproduction, biology of, 66
    in cattle, 77–80
    female reproductive system,
        69–71
    in horses, 81–82
    male reproductive organ,
        71–72
    reproductive cycle, 72
        fertilization, 72–73, 74
        lactation, 75
        parturition, 74–75
        pregnancy, 74
    in sheep, 80–81
    in swine, 76–77
Reproductive diseases, of beef
    cattle, 185, 186
Retail cuts, 338
Retail value, 336
Retained placenta, 216
Rete testis, 71
Reticulorumen, 62, 63
Ribosomes, 6
Rigor mortis, 45
Ringworm, 184

Roan Shorhorns, 91
Roasting, 338
Robotic milking systems, 218
Rodeo horses, 261, 289
Rolled oats, 244
Rotaterminal crossbreeding,
    131–132
Rotational grazing, 221
Rumen, 52, 53, 62
Rumen fermentation, 212
Rumen-protected fat, 212
Ruminants, 54, 62

S

S period, 6, 7
S-shaped growth curve, 23
Salers, 174–175
Salmonella, 186
Salt, 279
Sample
    beef carcass, 353–354
    lamb carcass, 363
    pork carcass, 359
Sarcomeres, 42
Sarcoplasmic reticulum (SR), 43
Satellite cells, 28–29
Scenarios, 324–327
    Angus Heifers, 326–327
    Limousin Bulls, 325–326
Scrapie, 253
Scur, occasional, 233
Seasonal breeders, 80, 233.
        See also Dorset
Seedstock, 151
Seedstock operations, 190
Segregated production, 138
Selection intensity, 97–99
Seminiferous tubules, 71
Sensors, 218
Settle, 178
Sex chromosomes, 89
Sexual behaviors, 107
Shearing, 245
Sheep industry, 228
    career opportunities, 254
    industry organization,
        251–252
    issues, 252–254
    marketing, 250–251

    supporting organizations,
        254–255
Sheep management, 228,
    244–245
    breeding evaluation, 398–402
    breeding systems, 238–240
    breeds, 232
        Columbia, 235, 236
        Corriedale, 237
        Dorset, 233–234
        Hampshire, 234
        Polypay, 235, 236
        Rambouillet, 234–235
        Shropshire, 238
        Southdown, 237
        Suffolk, 232–233
        Texel, 238
    diseases, 247
        Clostridia, 247–248
        foot rot, 248
        pneumonia, 248
        sore mouth, 248–249
    housing, 249–250
    nutrition, 240–244
    parasites
        coccidiosis, 247
        worms, 246–247
    taxonomy of, 16
Shipping fever, 185
Shorthorns, 165
Shropshires, 238
Sickle-hocked, 372, 373
Silage, 62, 211
Simmental cattle, 175
Simmering, 338
Simple carbohydrates, 56
Sire breeds, 121
Sister chromatids, 7, 8
Skeletal muscle, 27
Skillathon contests, 308
Sleeping sickness, 283
Slow, red fibers, 44, 45
Small-framed cattle, 182
Smooth muscle, 27
Social behaviors, 104, 106–109
Sodium bicarbonate, 212
Somatic cell count, 219
Somatotropin (ST), 32–33
Sore mouth, 248–249
Southdown, 237

Sow productivity index (SPI), 320

Soybean meal, 244, 277

Spanish Merino breed, 234

Species, 13

Splay-footed, 372, 373

"the split-eared heifer", 381

Split-sex feeding, 140

Spool joints, 361

Spots. *See* Spotted Swine

Spotted Swine, 126, 127

Stanchion barns, 216

Standardbreds, 266

Standing heat, 70

Steers, 24

Steps involved in yield grading, 339–341

Stillborn (born dead), 246

Stocker operations. *See* Grower operations

Stomach worms, 185

Strangles. *See* Distemper

Streptococcus, 285

Stretch-induced hypertrophy, 41

Subcutaneous fat, 29, 30

Subprimal cuts, 337

Suffolks, 232

Sulky, 262

Sus scrofa, 14

Swine
   breeding evaluation, 394–398
   market evaluation, 376–383
   underline faults in, 395

Swine erysipelas, 143

Swine industry, 118
   career opportunities, 153–155
   industry issues, 151–153
   industry organization, 151, 152
   marketing, 148
      cash price, 150
      formula price, 151
      futures price, 150–151
   overview, 120–121
   supporting organizations, 155–156

Swine management, 118
   breeding systems, 130
      backcrossing, 130
      rotaterminal crossbreeding, 131–132

terminal crossbreeding, 132–133
      three-breed rotational crossbreeding, 131
   breeds, 121
      Berkshire, 122–123
      Chester White, 123
      Chinese strains, of swine, 129
      Duroc, 123
      Hampshire, 124, 125
      Herefords, 128
      Jersey Reds, 123–124
      Landrace, 125
      Pietrains, 128, 129
      Poland Chinas, 126
      Spotted Swine, 126, 127
      Tamworth, 128
   diseases, 142
      atrophic rhinitis, 143
      leptospirosis, 143–144
      mycoplasmal pneumonia, 144
      parvovirus, 144
      PRRS (porcine reproductive and respiratory syndrome), 145
      pseudorabies (PRV), 145
      swine erysipelas, 143
   housing, 145–148
   nutrition, 138–141
   parasites, 141
      lice, 142
      mange, 142
      worms, 141–142
   piglet and nursery management, 135–138
   swine production cycle, 133–135
   taxonomy of, 14–15

Swine production cycle, 133–135

Symbiotic relationship, 64

Synthetic breed, 235. *See also* Polypay

T

Tack, 288

Tail docking, 245

Tamworth, 128

Teasing, 81

Telophase, 8, 10

Tenderness, 319

Tennessee Walker. *See* Tennessee Walking Horse

Tennessee Walking Horse, 267–268

Terminal crossbreeding, 132–133

Terminal sire, 132, 175

Terminal sire indexes (TSI), 320

Testis, 71

Testosterone, 71, 72

Tetanus, 286

Texel, 238

Thermoregulatory behaviors, 105

Thoroughbreds, 263

Three-breed rotational crossbreeding, 131

Tie-stall housing, 216

Titin, 42, 43

Tongue prehenders, 105

Transition phrase, 303

Trefoil, 277

Trichomoniasis, 186

Trimness, 338

Tropomyosin, 42, 43

Troponin, 42, 43

True stomach, 62

Tuberculosis (TB), 187

Turnips, 240

Twisted stomach, 215

Type classification, 409–413

Type-production index (TPI), 322

U

Uncastrated, 27

Underfinished cattle, 371

Undulent fever, 186

Unfermented forages, 277

Unfinished cattle, 371

United Dairy Industry Association (UDIA), 224

United States Department
of Agriculture (USDA)
meat inspectors,
332
United States Equestrian
Federation, 290
Unregistered animals, 205
U.S. Beef Breeds Council, 193
USDA meat graders, 333
USDA graders, 338
Uterus, 70, 71, 77

V

Vagina, 71
Vas deferens, 72
Veal production, 205
Villi, 59
Vitamins, 54–55, 279
Vitelline block, 73

Volatile fatty acids (VFAs), 63
Vulva, 71

W

Water, 53
Wean to finish system, 138
Weaning, 178
Weaning weight EPD (WW
EPD), 321
Well-marbled cattle, 188
West Nile Virus, 286
Western range ewes, 239
Western range flocks, 240,
251, 255
Wethers, 27
Wheat, 277
Winter, 190
Women's Professional Rodeo
Association, 290

Wool clip, 255
Wool grading systems, 251
Wool, market price of, 251
Worms, 246–247

Y

Yearling breeding, 277
Yearling weight EPD (YW
EPD), 321
Yorks. *See* Yorkshires
Young timothy hay, 277

Z

Z-lines, 42
Zona pellucida, 73
Zygote, 73

DATE DUE